IMAGES
of America

FIREFIGHTING IN HAGERSTOWN

UNITED FIRE-COMPANY

WE the ſubſcribers, inhabitants of *Elizabeth-Town*, taking into conſideration the ruinous conſequences of the fierce and unrelenting element of FIRE; in order to prevent as much as poſſible the waſte and devaſtation which may be occaſioned thereby, have determined to yield and afford our United Strength; for which purpoſe we form, agree to, and ſubſcribe the following Conſtitution.

I. This inſtitution ſhall be known by the name of the UNITED FIRE COMPANY.

II. That each member of the Company (not already furniſhed therewith) ſhall provide for himſelf two or more leathern buckets and one bag of good oznaburgs, or wide ſtrong linen, containing not leſs than two and an half yards, with a running ſtring at the mouth; which buckets and bag ſhall be marked with the name of the member to whom they reſpectively belong, and recorded in the Company's books, and ſhall be kept in ſome convenient place of each member's houſe, and applied to no other uſe than what is hereby intended.

III. That the Company ſhall appoint one of their members to act as Clerk for twelve months, at the end of which time he ſhall not be obliged to act as Clerk again until the end of the ſecond year following; and the Clerk for the time being who ſhall neglect to officiate in perſon, as directed, to enter the proceedings of the Company, or who ſhall, in any caſe of abſence, neglect to procure another member of the Company to act for him, he ſhall for every ſuch neglect forfeit and pay the ſum of *Three ſhillings and nine pence, (Fifty Cents)* provided, if it ſhall appear that ſuch Clerk's abſence is occaſioned by ſickneſs, and he ſends the books of the Company to the meeting, he ſhall be exempt from this fine.

IV. That the Company ſhall meet on the firſt Saturday in January, April, July and October, at 4 o'clock P. M. at the Court-Houſe, and the members preſent (not leſs than twenty) ſhall conſtitute a quorum to do any buſineſs of the Company, except altering this Conſtitution, and may, at their diſcretion, adjourn to any other place for the conveniency of doing buſineſs; and every member neglecting ſo to meet, ſhall for every ſuch neglect, forfeit and pay *One ſhilling and ſix pence, (Twenty Cents)* unleſs it ſhall appear that ſuch member's abſence is occaſioned by ſickneſs.

V. That at every meeting of this Company a Preſident ſhall be appointed by a majority of the members preſent, before any other buſineſs is done, who is empowered to keep order and all motions ſhall be addreſſed to him; and every member who makes a motion ſhall riſe at the time; and any motion made and not ſeconded within a reaſonable time ſhall be loſt at the diſcretion of the Preſident.

VI. That each member be furniſhed with his buckets and bag within three months after being admitted as a member of the Fire Company, and until the expiration of that time, no fines ſhall be inflicted for neglecting to have them and keeping them in good order. And in caſe any member ſhall loſe a bucket, buckets, or bag at any Fire, the ſame ſhall (on notice thereof) be replaced to ſuch member out of the Company's ſtock, provided the ſame be recorded.

VII. The Company ſhall appoint annually one of their members to ſerve as their Treaſurer, who ſhall give his bond, for the uſe of the Company, to two other members who may be appointed for that purpoſe, in a penalty not leſs than double the ſtock then in hand, for the faithful performance of the truſt repoſed in him, and for the d[illegible]ment of ſuch ſtock, or any part thereof, in ſuch manner as the Company may b[illegible] order direct; and the Treaſurer's bond ſhall be renewed at every appointment.

VIII. That twelve members be appointed quarterly, to examine the buckets & bags of the company, monthly, four at a time, in rotation, and report thereon, at the next meeting, and for every neglect, ſuch member ſo appointed ſhall forfeit and pay the ſum of *Three ſhillings and nine pence, (Fifty Cents)* that every member ſo appointed, who ſhall neglect any part of ſuch duty required, and report all in good order, upon conviction of ſuch neglect, by one or more members of the Company, ſhall be fined *One Dollar*, for ſuch neglect; and further, that the Clerk for the time being, ſhall give notice in writing, to any one of the four members appointed for each month in rotation as viſitors, at every meeting, and before the Company adjourns, or as ſoon after as poſſible, the names of the viſitors ſo appointed, the member receiving ſuch notice and neglecting to notify the others or any of them, as is hereby intended, ſhall be fined *One Dollar;* but if no notice be given by the Clerk, then there ſhall be no fine—and for every bucket and bag found out of order or not in the proper place at the time of viſiting, the reſpective owner thereof ſhall forfeit and pay the ſum of *One ſhilling and ſix pence. (Twenty Cents)* for each default, and it ſhall be the duty of the reviewers to collect all fines, forfeitures and penalties impoſed by this Conſtitution, and pay the ſame to the Clerk of the Company, who ſhall enter the ſame in the Company's Books, and deliver it over to the Treaſurer for the time being.

IX. That no perſon ſhall hereafter be admitted a member of this Company without leave of the ſame. The time of calling the roll and entering on buſineſs ſhall be *Fifteen minutes* after the time appointed to meet. No member ſhall leave the Company before the buſineſs is done and the meeting adjourned (without leave of the Preſident) under the penalty of *Two ſhillings and ſix pence—(Thirty three and one third Cents)*

X. That each member ſhall affix in his Houſe, in open view and near his buckets and bag, a printed copy of theſe articles with a liſt of the members' names, and in default thereof ſhall forfeit and pay the ſum of *One ſhilling, (Thirteen and a half Cents)* at each review. And any member having a deſire at any time hereafter to withdraw himſelf from this Company, on application, and on relinquiſhing his claim and right to any part of the Company's ſtock, they may, at their diſcretion, diſcharge ſuch applicant.

XI. That it ſhall be the duty of each member to have affixed on the upper part of the front Door frame of his Houſe, a plate of Copper, Braſs or Tin, of an Oval form, five inches in length, painted black, and marked with white letters, U. F. Co. within ſix months—and, in caſe there ſhall be any new member admitted, he ſhall provide himſelf as aforeſaid, within three months after ſuch admiſſion, and in caſe of neglect he ſhall be fined *One ſhilling and ſix pence—(Twenty Cents.)*

XII. That if any member ſhall diſregard theſe articles, and refuſe to pay the ſeveral fines, forfeitures and penalties impoſed, when due, we and each of us reſpectively, authoriſe any Juſtice of the Peace, on application of the Company, ſigned by their Clerk, to iſſue proceſs againſt us, and each of us reſpectively, thus making default, our and each of our Heirs, Executors or Adminiſtrators, and to award judgment, and iſſue execution thereon as in caſes of ſmall debts.

XIII. That any other rules or alterations in this Conſtitution, which may be entered into or made by this Company, when a majority of the members are preſent, ſhall be deemed as binding to all intents and purpoſes as thoſe now formed and ſubſcribed to.

IN WITNESS whereof we have hereunto ſeverally ſigned and ſubſcribed our names this fifth day of February, 1803.

1 Adam Ott
2 William Lee
3 Richard Pindell
4 John Lee
5 John Geiger
6 John Ragan
7 David Harry
8 Jacob Schnebly
9 Jacob Harry
10 George Woltz
11 Daniel Nead
12 Robert Douglaſs
13 Frederick Alder
14 George Shall
15 Samuel Young
16 George Shank
17 Nathaniel Rocheſter
18 Chriſtopher Alder
19 Alexander Clagett
20 Charles Gelwicks
21 William Heſs
22 Henry Cake
23 John Protzman
24 Michael Kapp
25 Peter Hœflich
26 Conrad Coffroth
27 Otho H. Williams
28 Peter Miller, (Hatter)
29 Peter Woltz
30 William Clagett
31 William Heyſer
32 Thomas Grieves
33 Peter Miller, (Merchant)
34 Henry Middlecalf
35 James Ferguſon
36 Leonard Kuhn
37 John Herſhey
38 John Wiſe
39 Daniel Heiſter
40 Benjamin Clagett
41 Chriſtian Hawken
42 George Binckley
43 Philip Kellar
44 [illegible]am Kreps
45 John Miller
46 John Reynolds
47 Jacob D. Dietrick
48 Chriſtian Langenecker
49 Alexander Neill
50 Timothy Monahan
51 John Cook
52 Peter Umrickhouſe
53 Samuel Beelor
54 John Freaner
55 George Kreps
56 Henry W. Lewis
57 John P. Herr
58 Seth Lane
59 Jonathan Rahauſer
60 Adam Borhoff
61 Martin Kreps
62 John Conrad
63 John Heddinger
64 Frederick Miller
65 Jonathan Hager
66 Devalt Eichelberner
67 George Criſſinger
68 Frederick Dorſey
69 John Julius
70 Benjamin Light
71 David Ridenour
72 George Hammer
73 Samuel Hughes
74 Baltzar Bowman
75 Henry Snyder
76 John A. Donaldſon
77 Chriſtian Stemple
78 Abraham King
79 Henry Burkman
80 Jacob King
81 John Sleigh
82 Jacob Kraft
83 Henry Kaelhofer
84 John Kaufler
85 Henry Dillman
86 Chriſtian Fechtig
87 Henry Strauſe
88 Jacob Binckley
89 William Reynolds
90 Charles Oldwine
91 George Martini
92 Lewis Blaire

Prior to 1791 there was no organized fire company in Hagerstown. When a fire occurred, neighbors would gather around with as many buckets and tubs as they could secure. At a meeting at the courthouse on New Year's Day 1791, the United Fire Company was formed. On February 5, 1803, this constitution was adopted. At the time, the town was called Elizabethtown, named after the wife of the town's founder, Jonathan Hager. One prominent members, Nathaniel Rochester, was the founder of Rochester, New York. Note the old English spelling. Most of the *F*s are *S*s in today's spelling. (Courtesy of the James Plummer collection.)

IMAGES
of America

FIREFIGHTING IN HAGERSTOWN

Justin T. Mayhue

ARCADIA
PUBLISHING

ISBN 978-0-7385-1658-5

Published by Arcadia Publishing
Charleston, South Carolina

Printed in the United States of America

Library of Congress Catalog Card Number: 2004103640

For all general information contact Arcadia Publishing at:
Telephone 843-853-2070
Fax 843-853-0044
E-mail sales@arcadiapublishing.com
For customer service and orders:
Toll-Free 1-888-313-2665

Visit us on the Internet at www.arcadiapublishing.com

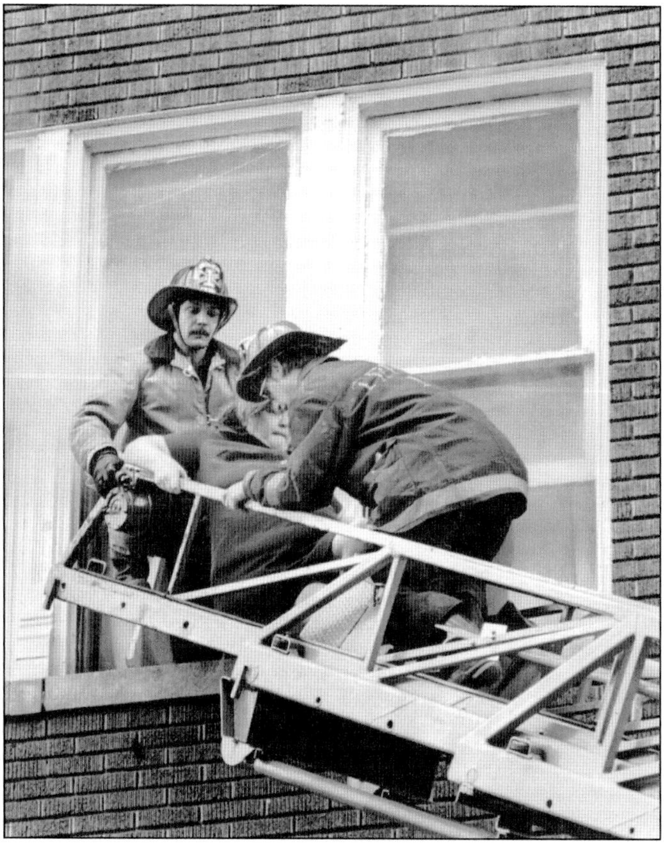

On February 1, 1977, box alarm 625 was sounded for a building fire at the Glick Shoe Store located in the unit block of West Washington Street. Ultimately, this would become a three-alarm fire. At least four citizens were rescued by firefighters. This citizen is being rescued by members of the Pioneer Hook & Ladder Company, Truck 1, Fire Apparatus Operator (FAO) Rob Dieterich (left) and Firefighter Glen Giffin. (Copyright *The Herald-Mail* Company, Hagerstown, Maryland.)

Contents

Dedication

The following died in the line of duty:
John Fridinger of the Independent Junior Fire Company died on December 6, 1871. The cupola and roof fell on him while fighting a fire at the Washington County Courthouse, located at the intersection of West Washington Street and Summit Avenue.
William P. Negley, Pioneer Hook & Ladder Company, died September 19, 1930, from injuries received after being struck by a vehicle while on duty near the intersection of Franklin and Potomac Streets.
Walter Reige Sharar, Antietam Fire Company, died May 23, 1932, from injuries received in a fall from a ladder while fighting a fire at 181 South Prospect Street.
Frederick W. White, Antietam Fire Company, died March 16, 1988, from a heart attack while on duty at the Antietam Fire Company at 113 Summit Avenue.

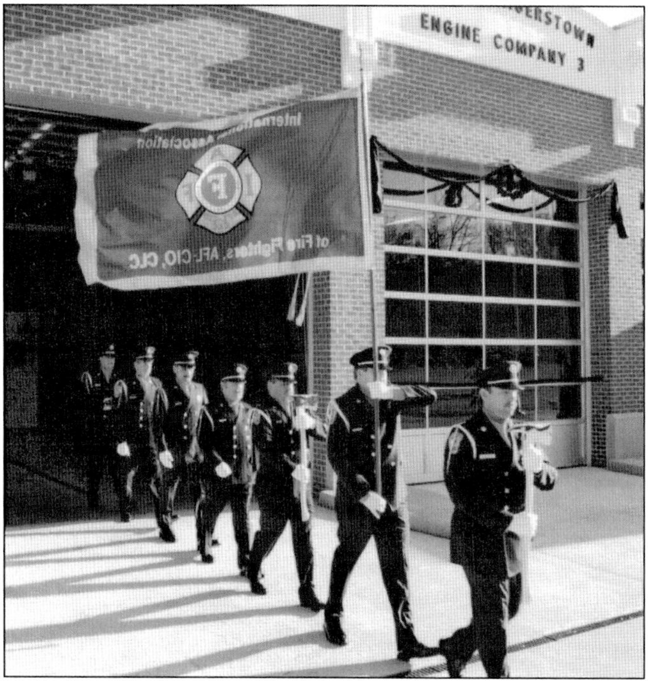

The International Association of Firefighters (IAFF) Local 1605 Honor Guard provides support and comfort to a fallen firefighter's family. The honor guard also participates in parades and other formal ceremonies. (Courtesy of the C. Edward Shindle collection.)

INTRODUCTION

The Hagerstown Fire Department is rich in history. The successes of our forefathers determine our future. Until now, no complete photographic history of the 213 years of firefighting in Hagerstown has ever been compiled. Past generations deserve to be remembered for their accomplishments.

The United Fire Company was in existence until about 1814. In 1817, a group of citizens, some of whom were past members of the United Fire Company, formed the Hagerstown Fire Company. By 1824, the following three fire companies were in existence: the First Hagerstown Hose Company, which was incorporated in January 1823; the Washington Mechanic Fire Company; and the Franklin Union Fire Company. In the beginning of 1842, the three fire companies were the First Hagerstown Hose Company, the Antietam Fire Company, and the Franklin Union Fire Company. Early in 1842 a number of the Franklin's young members convinced town officials to turn over the fire equipment to them in order to create the Independent Junior Fire Company. By 1872, the Western Enterprise Fire Company and the Pioneer Hook & Ladder Company were part of the Hagerstown Fire Department. In 1894, because part of the expanding town was west of the Cumberland Valley Railroad, the Columbian Chemical Company was formed. The Columbian's firehouse was located in the unit block of High Street. The Columbian Chemical Company was only in existence until about 1903. The most recent fire company in Hagerstown is the South Hagerstown Fire Company, which began responding to alarms in 1950.

In the early days, the fire apparatus was housed in sheds or garages. The company meetings were usually held at the town hall. But by the end of the 1800s, the firehouses or fire halls were more than just a place to house the fire apparatus. With their huge size and elaborate facades, these buildings became the cornerstone of the community. Memberships swelled into the thousands. The fire companies held fairs, bazaars, conventions, and dances in grand ballrooms on upper floors. Space was also available for rental property next door to the apparatus bay. This provided income for the fire companies.

The citizens were requested to put candles in their windows at night upon hearing the cry of fire. In addition, the owner of every dwelling house with a value in excess of $300 was required to furnish two well-made leather buckets to be accounted for at all times. One leather bucket was required if the value of the dwelling was less than $300. The fine for not complying with the law was $1 per month. The firefighters were fined for various infractions within the fire company. If equipment was removed without authorization the fine was 50¢; 25¢ went to the company and 25¢ went to the informer. A fine of 50¢ was also imposed if the firefighter failed to respond to alarms.

The fire apparatus in Hagerstown has evolved over the years. In the beginning, bucket brigades were set up by lane directors who would direct two lines of men and women to shuttle water from a cistern or water pump to the fire. The first apparatus included hand-pumpers accompanied by a hose reel. By the late 1800s, the size and weight of the hose wagons and steamers necessitated the use of horses. Beginning in the early 1900s horses were replaced by motorized apparatus. Some of the manufacturers include Ahrens-Fox, Seagrave, American LaFrance, Knox, Pirsch, Mack, Robinson, KME, and Quality. Great effort and thought went into the purchase of each piece of fire apparatus, and a vehicle was expected to last between 20 and 30 years. Every effort has been made to include in this publication all motorized pumpers and ladder trucks that Hagerstown has ever owned.

Since about 1895 career firefighters/drivers have driven the fire apparatus. At first the individual volunteer fire companies paid the drivers, but sometime before 1910 the city began paying them. The term "driver" was used for career firefighters until the Union Local 1605 was initiated in 1966. Afterwards, the term used for career firefighters was Fire Apparatus Operator (FAO).

A piece of fire apparatus that contains a pump, water, and hose will be referred to interchangeably as a "pumper" or engine. A ladder truck or truck carries ground ladders and has an aerial device. A service truck has ground ladders but no aerial device.

I hope this photographic glimpse of the past will not only entertain but also serve as a lesson about where we came from and who we are today.

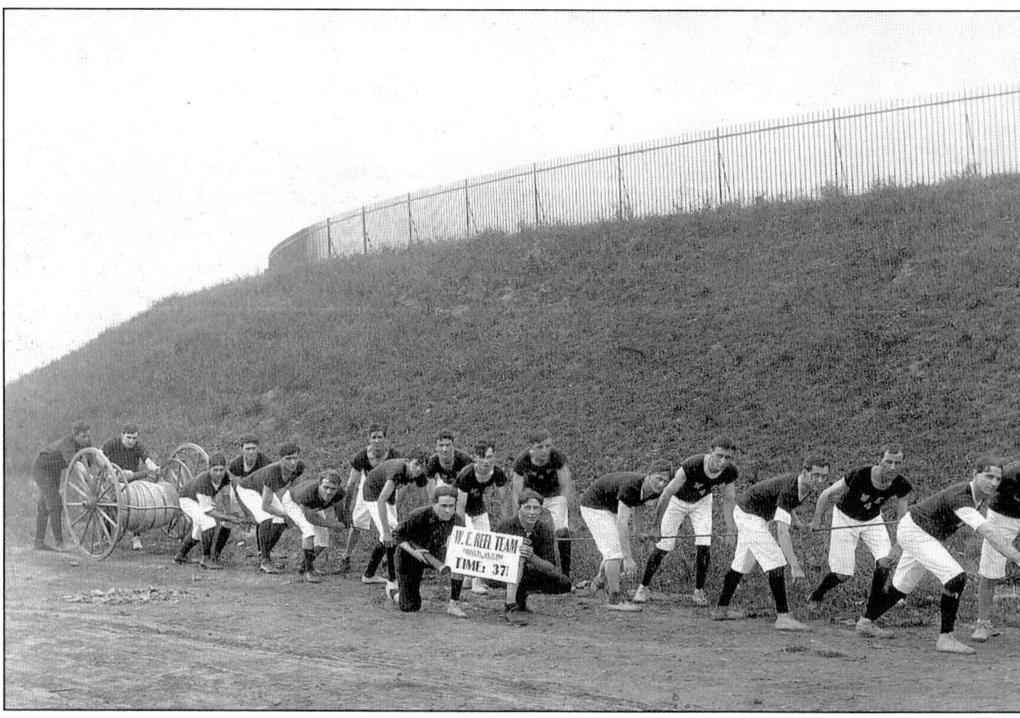

This is the Western Enterprise Hose Reel team after a competition at Carlisle, Pennsylvania, in 1909. (Courtesy of the Charles Cunningham collection.)

One

First Hagerstown Hose Company, Engine 1

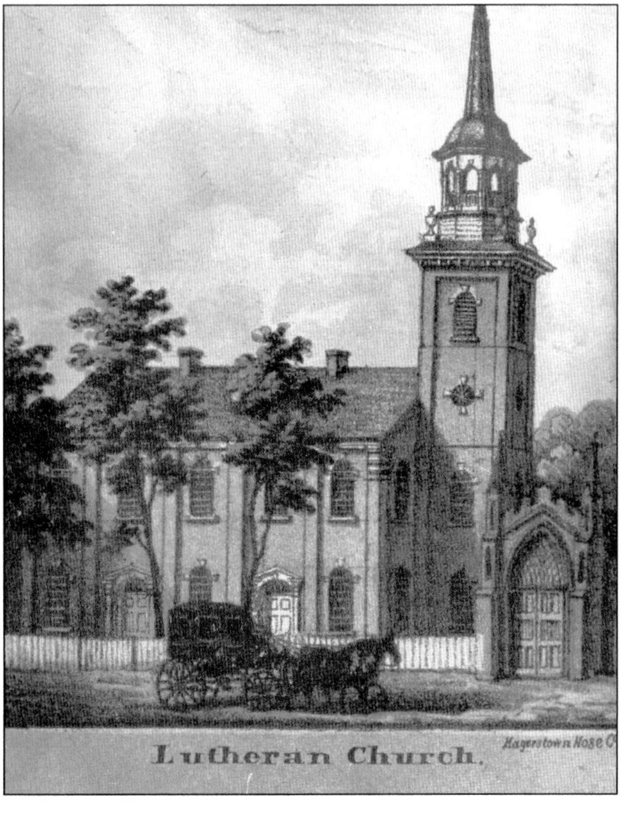

The First Hagerstown Hose Company (Engine 1) was organized as early as 1815. On January 8, 1823, the hose company was incorporated by an act of the Maryland General Assembly. The first engine house was situated upon the grounds of the Saint John's Lutheran Church at 141 South Potomac Street. The hand-pumper was housed in a small shed on the north side of the church. The frame shed was only large enough to house one piece of apparatus. The entrance to the shed was through the doors on the right. (Courtesy of the First Hagerstown Hose Company.)

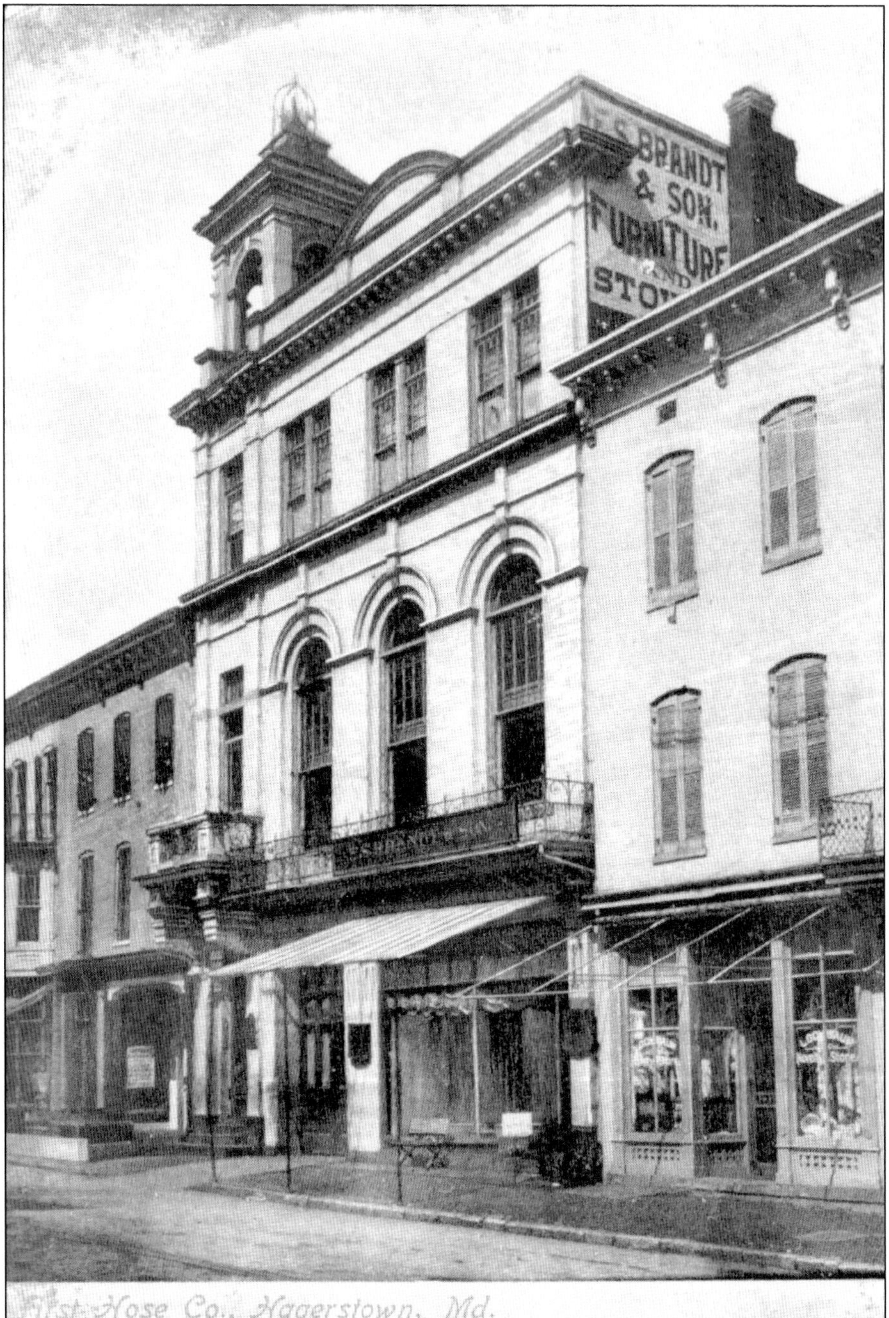

On January 27, 1881, the Hose Company purchased a lot from Daniel B. Stouffer at 31/33 South Potomac Street for $3,300 with the intent of erecting a permanent home. Local contractor Jacob Dayhoff received the contract to construct the building. The ground-breaking took place in July 1881. The total cost of the building was $15,310. Plans called for the building to be 100 feet by 40 feet and 3 stories tall. The building was named the Hose Opera House. Note the Brandt & Sons advertisement on the side of the building. The awning was added August 13, 1883. (Courtesy of the First Hagerstown Fire Company.)

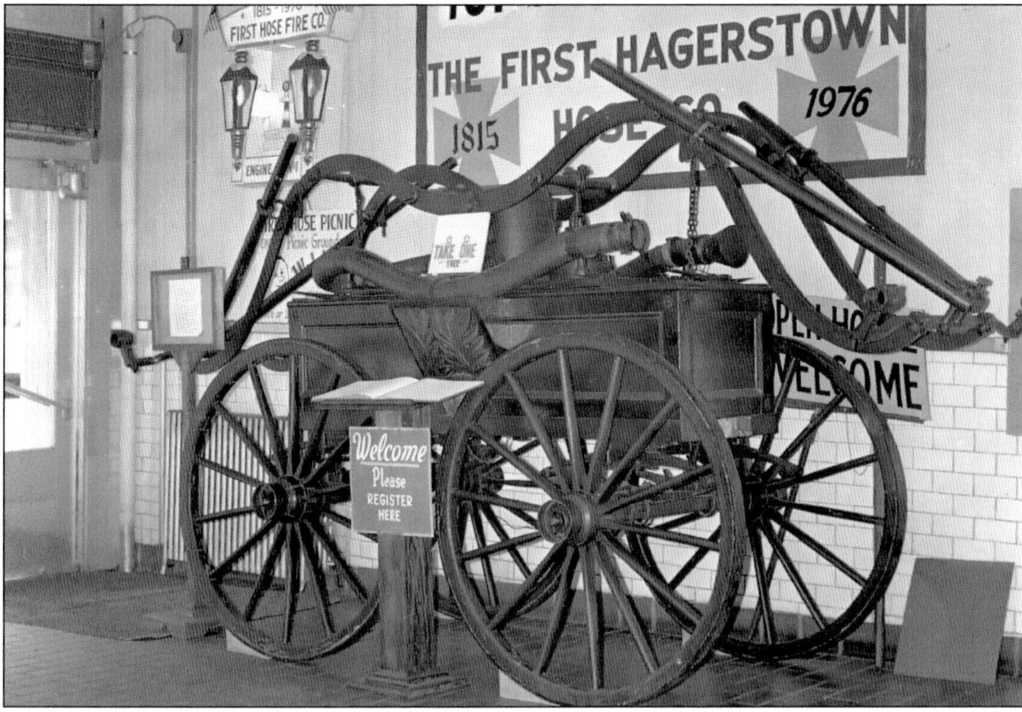

This is the oldest piece of fire apparatus in the city. It is an end-stroke hand-pumper made by the John Rogers Company of Baltimore, Maryland, in 1823. This hand-pumper carries the affectionate name of "Pet." "Pet" was small enough to reach cisterns in back alleys where larger apparatus could not fit. "Pet" is being displayed in the hose company engine bay. (Courtesy of the First Hagerstown Hose Company.)

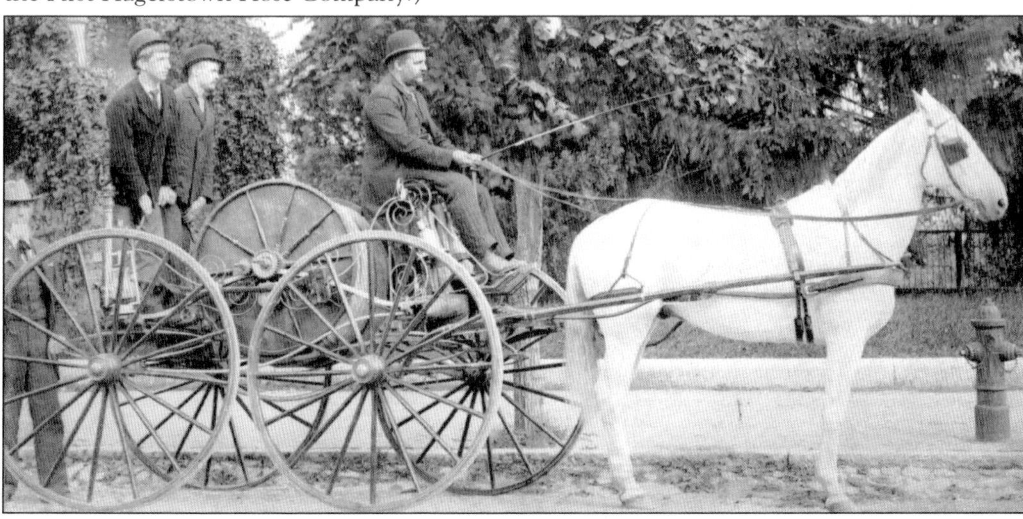

According to a letter, "Harry Bloom, owner of 'Grey Frank' and driver of the hose carriage for the hose company, was believed to be the first responder to Hagerstown fire on the dead run." The fire was at the Keeley Institute, located in the 500 block of Reynolds Avenue. The Keeley Institute was a franchise facility for the treatment of addiction to alcohol, nicotine, and narcotic drugs. The Keeley Cure relied heavily on injections of "bichloride" or "double chloride of gold" (hence the term "gold cure"). (Courtesy of the First Hagerstown Hose Company.)

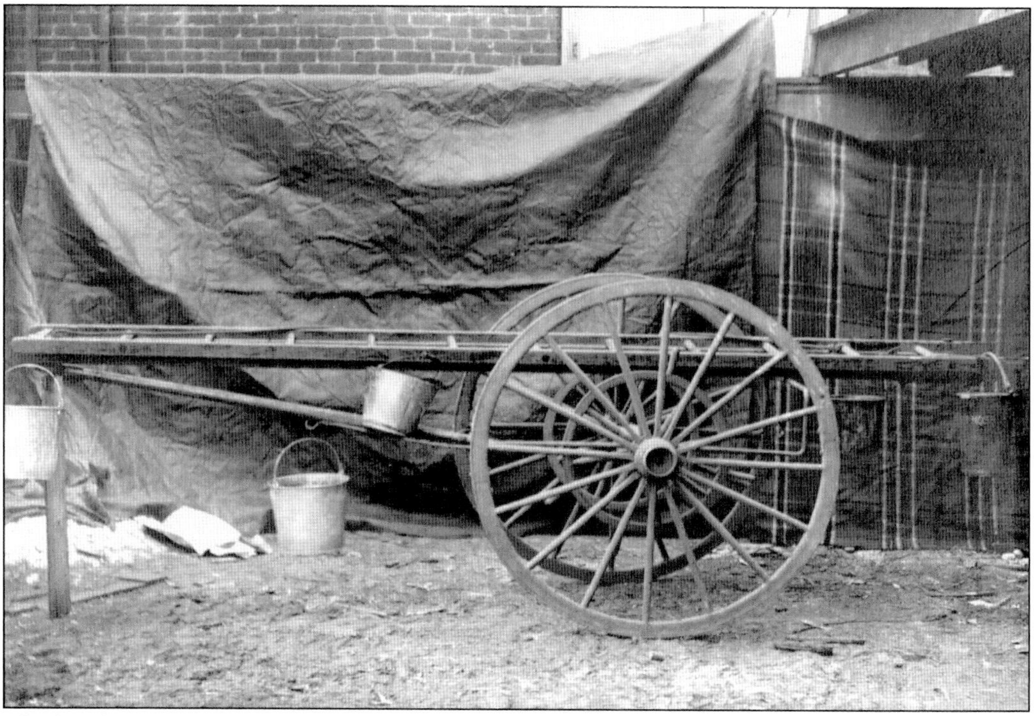

The hand-pumpers did not carry hose. A hand-pulled hose cart was used to transport the hose to the fire. This hose cart even has a ladder attached to the top. It was one of the hose company's first pieces of apparatus. (Courtesy of the First Hagerstown Fire Company.)

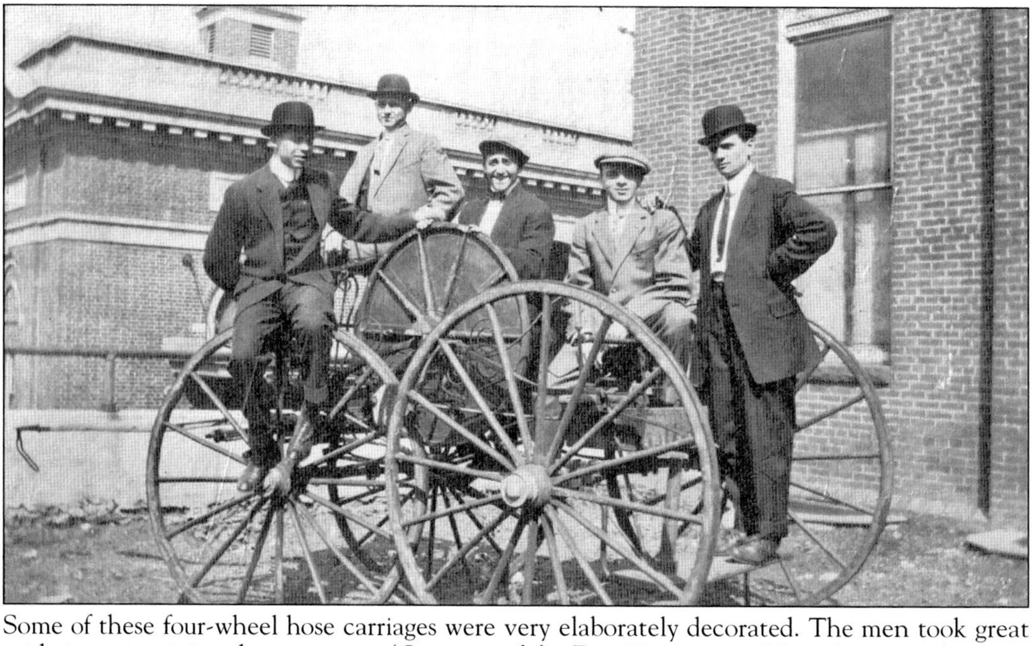

Some of these four-wheel hose carriages were very elaborately decorated. The men took great pride in maintaining the apparatus. (Courtesy of the First Hagerstown Hose Company.)

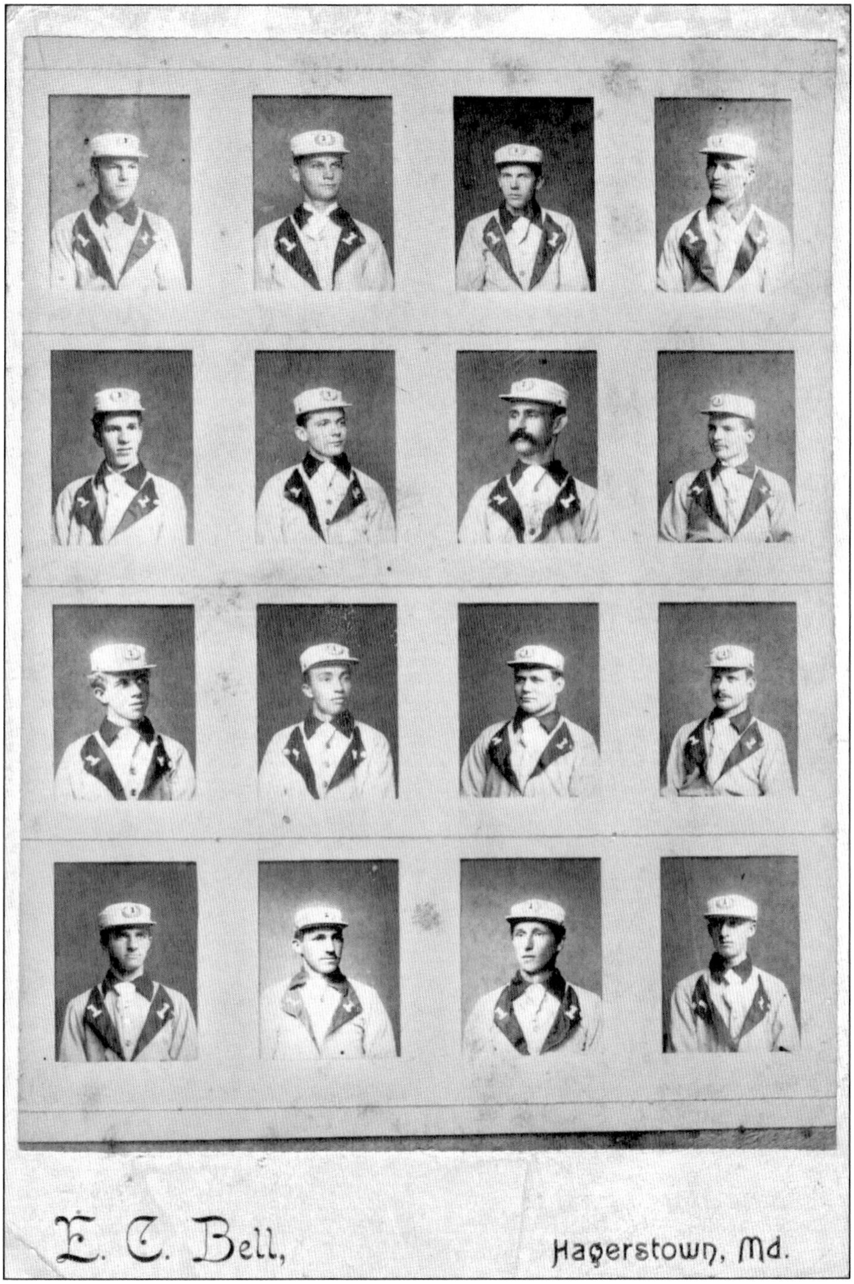

An important social function in the late 19th century and early 20th century was the reel races at the Maryland State Firemen's Association convention. These races were very competitive. Each fire company had to work as a team, just like in fighting a fire. The hose company team seen here took first place at the first annual convention held in Frederick, Maryland, on June 7 and 8, 1893. The members, from left to right, are (top row), Van Beachley, Guy Herman, Walter Smith, William Holland; (second row) Charles Roessner, Joe Hoffman, Emerich Bell, Joe Haeslip; (third row) Walter Byers, David Waltz, Augustus Heimel, Peter Brewer; (fourth row) Frank Middlekauff, Charles Hamburg, Andrew Smith, and Edward Baily. (Courtesy of the First Hagerstown Hose Company.)

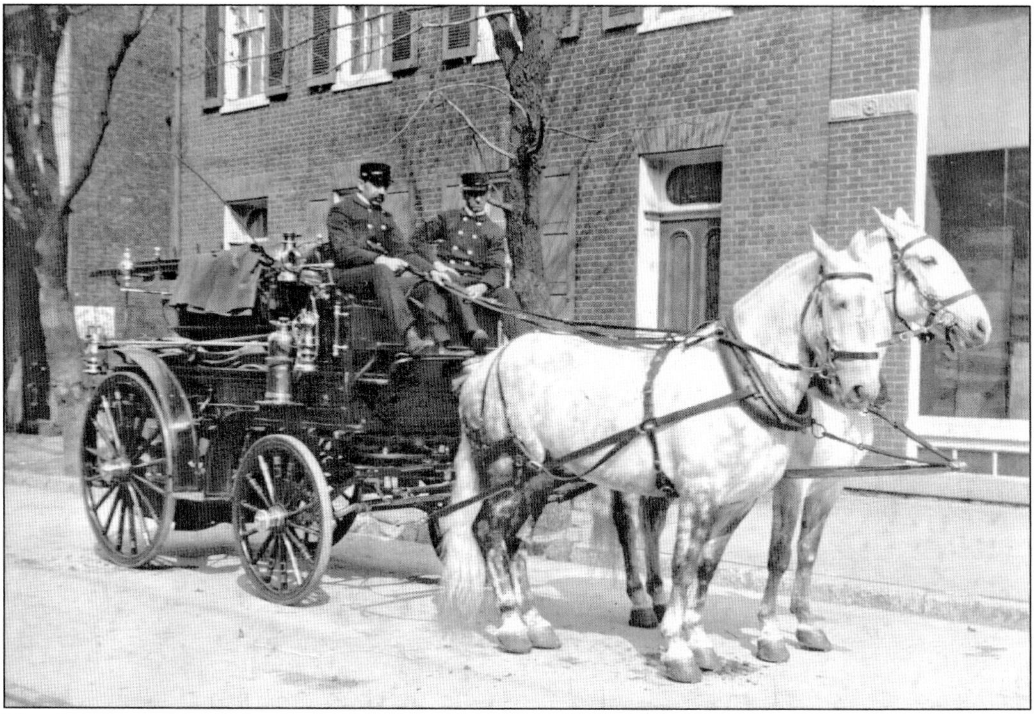

This hose wagon was purchased from the Rumsey Company of New York for the sum of $500 on January 7, 1895. The specifications included a wagon body to carry 1,000 feet of the heaviest-jacketed, cotton-rubber–lined, two-and-one-half-inch hose. The body was made of seasoned ash. The wagon weighed 1,800 pounds equipped. The horses are Rox and Don. This photograph was taken across the street from the hose company on South Potomac Street. (Courtesy of the First Hagerstown Hose Company.)

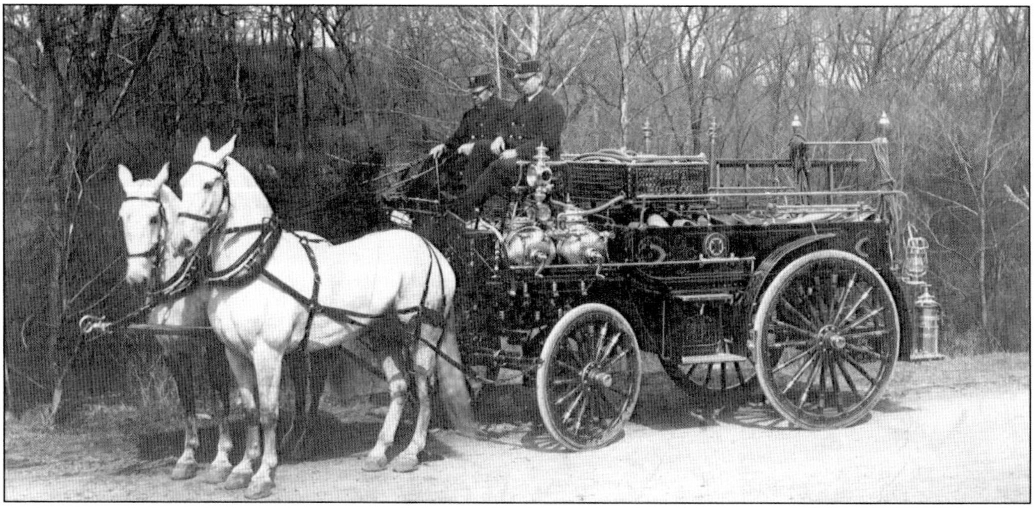

This combination hose wagon and chemical engine, pulled here by Rox and Don, was purchased from the Charles A. Holloway Company in 1903. The wagon was equipped with rubber wheels and a roller-bearing running gear. The roller bearing reduced the pulling weight by 40 percent. The wagon weighed 6,100 pounds and cost $2,500. (Courtesy of the First Hagerstown Hose Company.)

A fire dog is a man's and horse's best friend. One of the main reasons that dalmatians are known as fire dogs is because of the horses. The dogs were good companions and would protect the horses from other dogs. This photograph of Whitey and Spot was taken in March 1898. (Courtesy of the First Hagerstown Hose Company.)

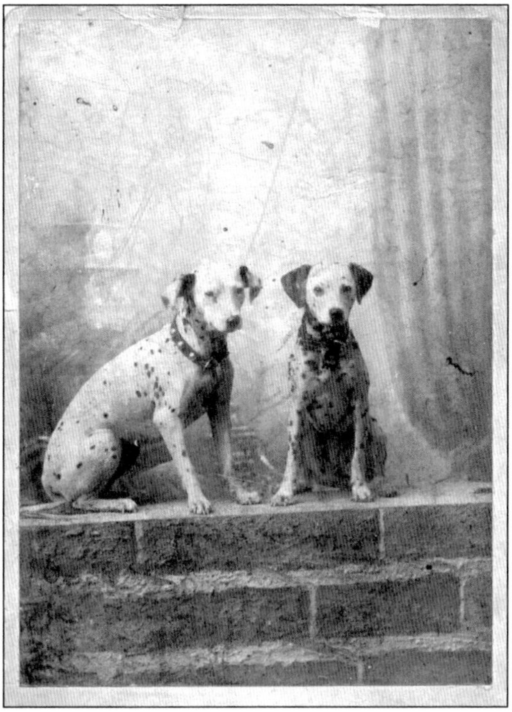

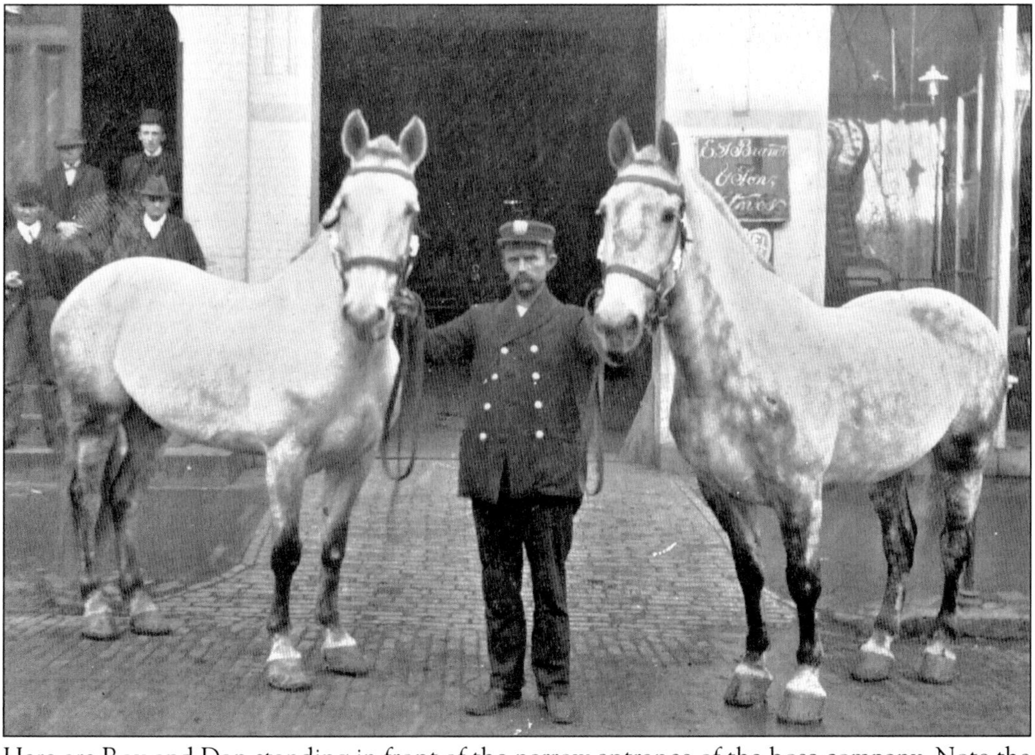

Here are Rox and Don standing in front of the narrow entrance of the hose company. Note the Brandt Furniture store rental property on the right and the entrance to the dance hall on the left. (Courtesy of the First Hagerstown Hose Company.)

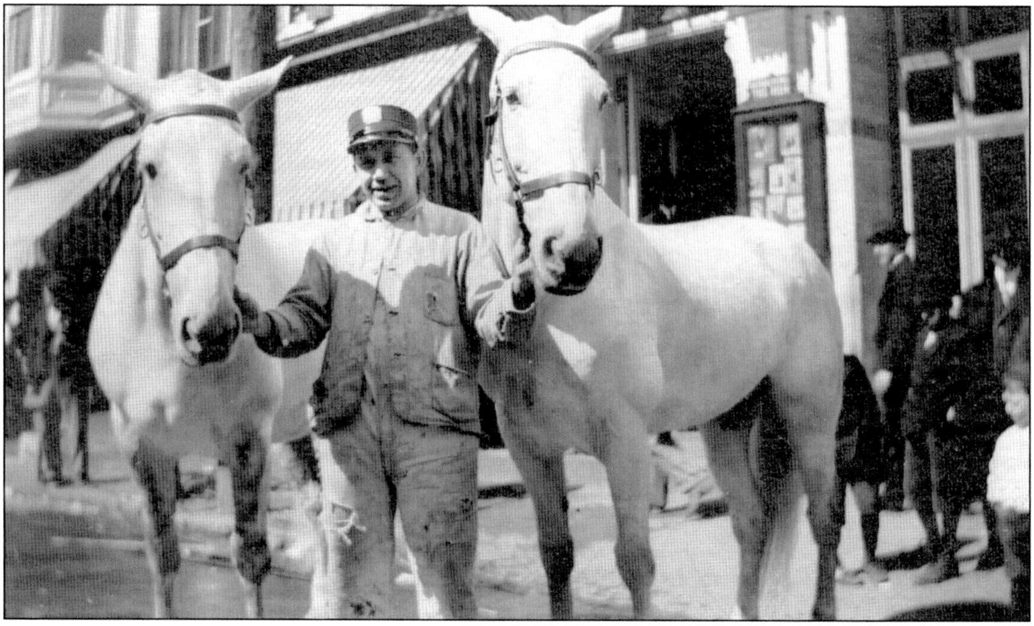

This is the retirement photograph of the prize-winning horses Rox and Don. It was taken in the spring of 1906. The horses are being held by the first paid driver of the hose company, Frank Maisack. Maisack started out as a volunteer firefighter and became a paid driver in 1895. Note Maisack's tattered work clothes and fine-looking stove-pipe cap. (Courtesy of the First Hagerstown Hose Company.)

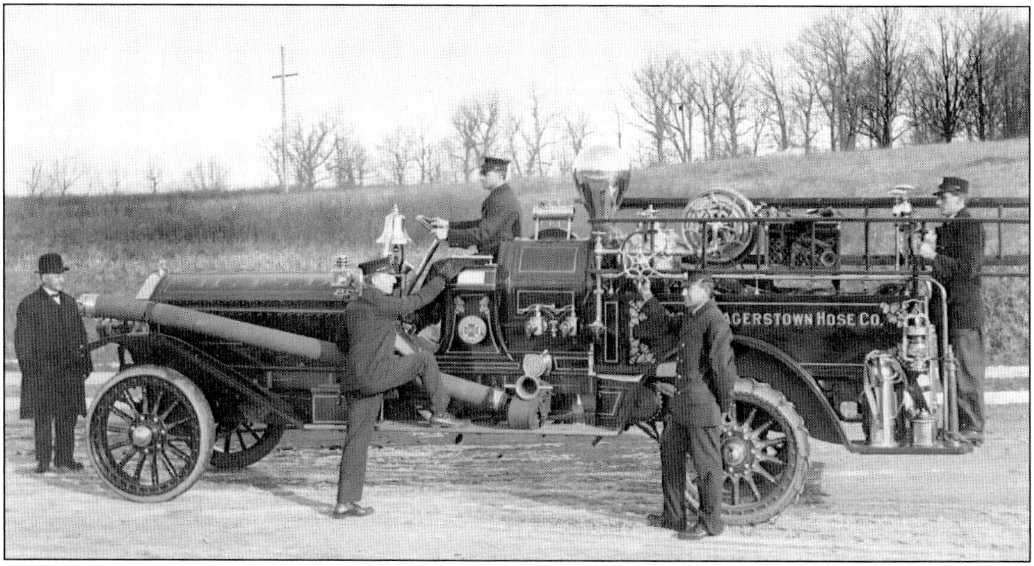

The hose company purchased this 1911 Robinson combination pumper in joint ownership with the City of Hagerstown. This was the hose company's first motorized equipment. The pumper had a 110-horsepower motor, 6-cylinder engine, 155-inch wheel base, and a top speed of 40 miles per hour. The pump was rated at 750 gallons per minute but during an acceptance test on January 4, 1912, the engine pumped 846 gallons per minute. Seated is the hose company's second career driver, Ben Leckron. Standing at the rear wheel is Frank Maisack. Standing on the rear tailboard is Scotty ?. (Courtesy of the First Hagerstown Hose Company.)

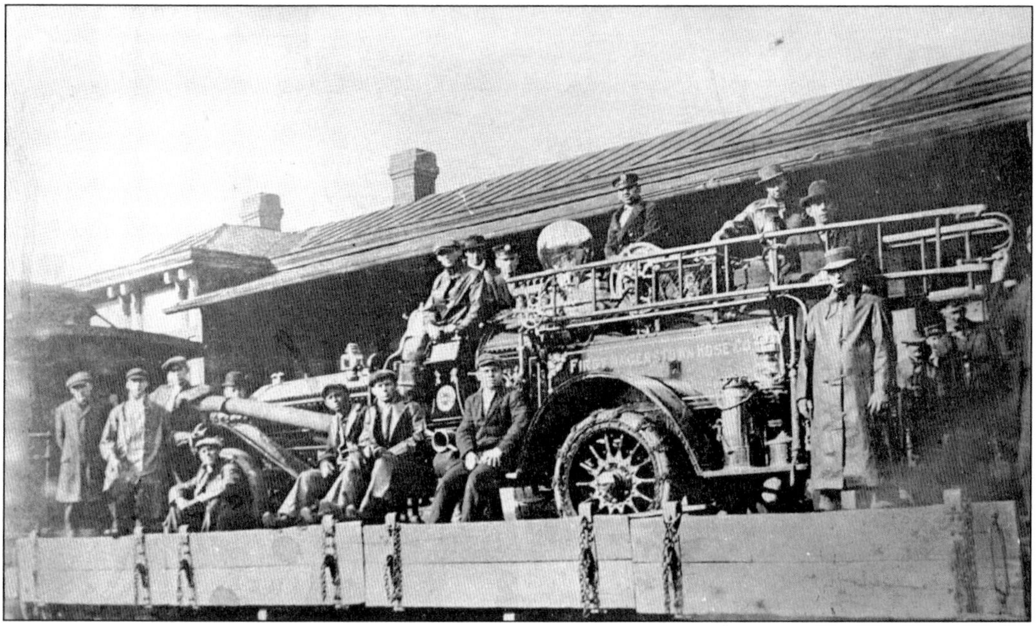

It was not uncommon in the early 20th century to transport fire apparatus by rail to fires in other towns. This is the 1911 Robinson responding to a fire in Shepherdstown, West Virginia, via the railroad. Note the chains on the solid rubber tires. (Courtesy of the First Hagerstown Hose Company.)

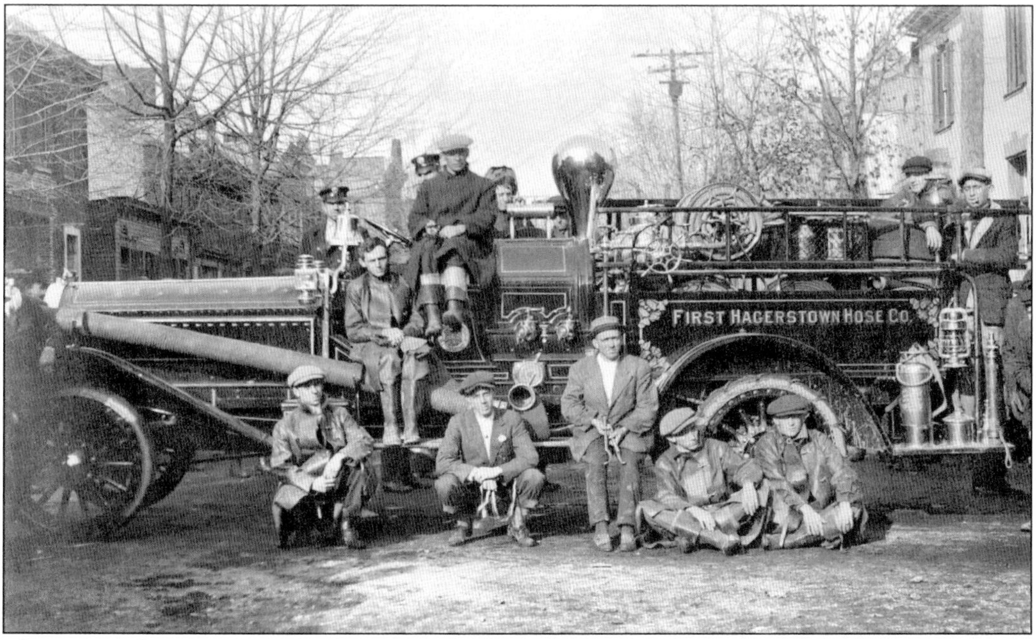

This group photo was taken after the fire in Shepherdstown, West Virginia. Note the protective equipment. The firefighters are wearing rubber coats and boots. There is no sign of helmets. The Robinson pumper had nickel trimmings and a 40-gallon chemical tank, and it cost $9,000. (Courtesy of the First Hagerstown Hose Company.)

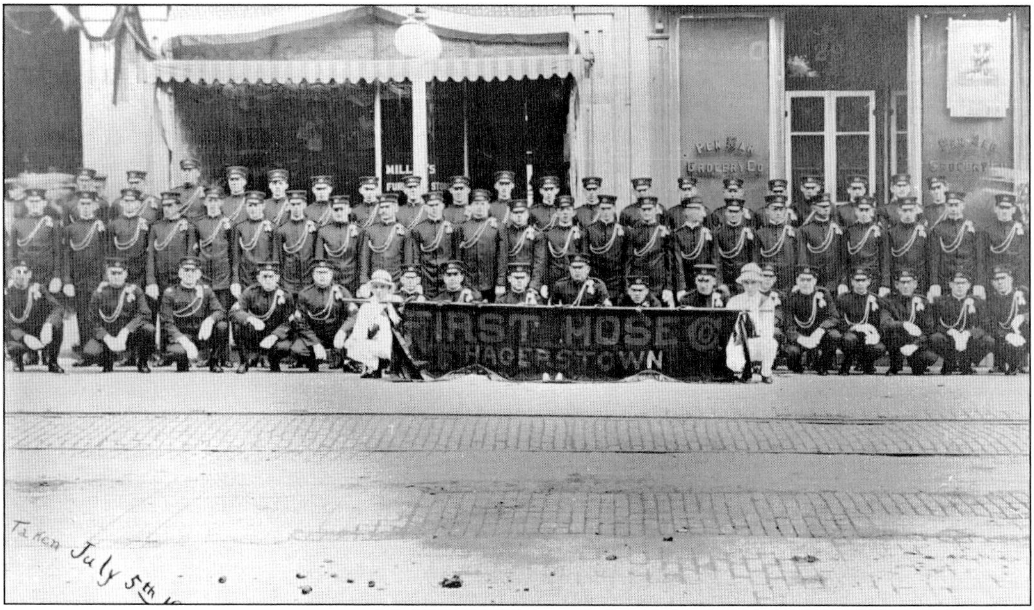

The active members of the First Hagerstown Hose Company are shown here in stylish parade uniforms. The photograph was taken in 1915 in front of the rental property occupied by the Miller Furniture Company. (Courtesy of the First Hagerstown Hose Company.)

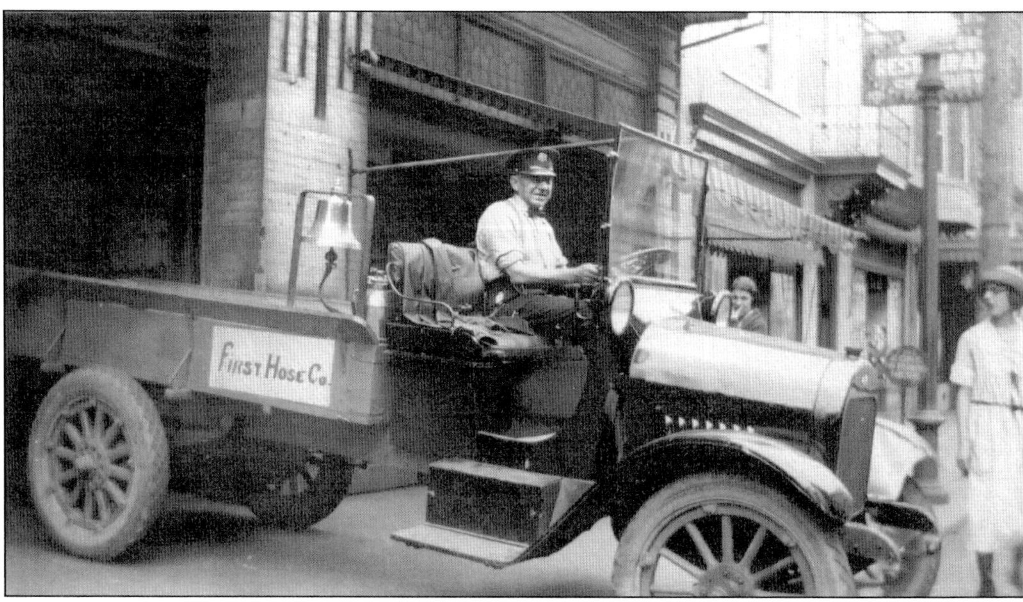

This Chevrolet truck served as a back-up piece of fire apparatus when the first due pumper was out of service. The man behind the wheel is Driver Frank Maisack. (Courtesy of the First Hagerstown Hose Company.)

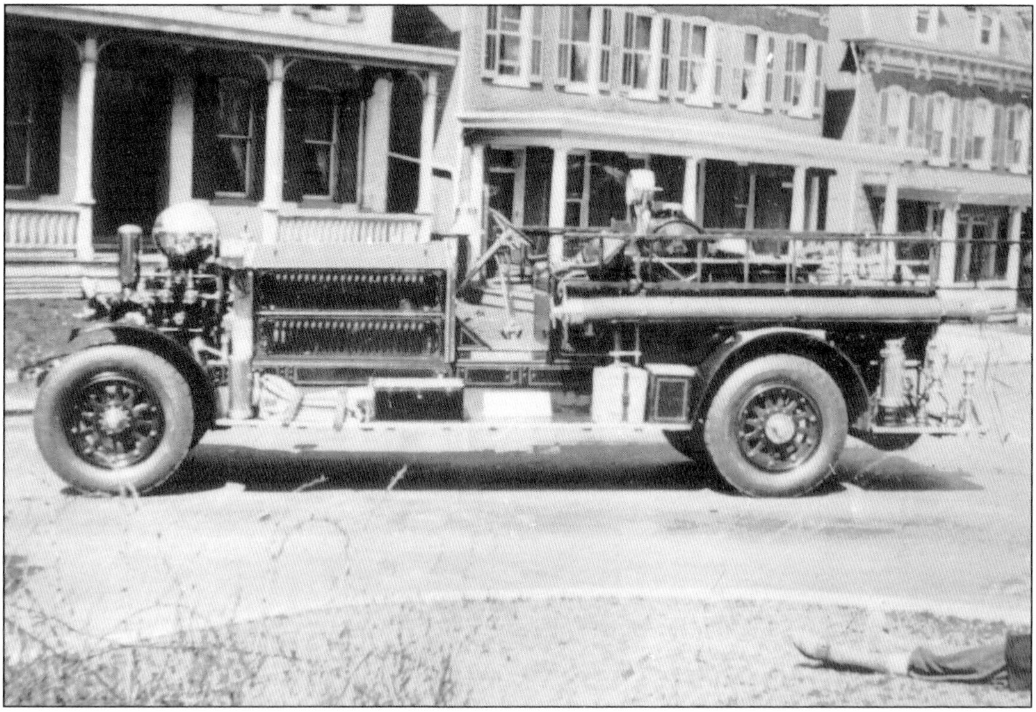

An Ahrens-Fox pumper can usually be identified by the big equalizing ball on the pump. This is an early photograph of the hose company's first Ahrens-Fox Model N-S-4 (Reg. No. 937) pumper. It is a right-hand drive, 1923 model with a 1,000-gallon-per-minute pump and 80-gallon water tank. The pumper is affectionately known as "Dynamite." Note the artillery-type spoke wheels. (Courtesy of the First Hagerstown Hose Company.)

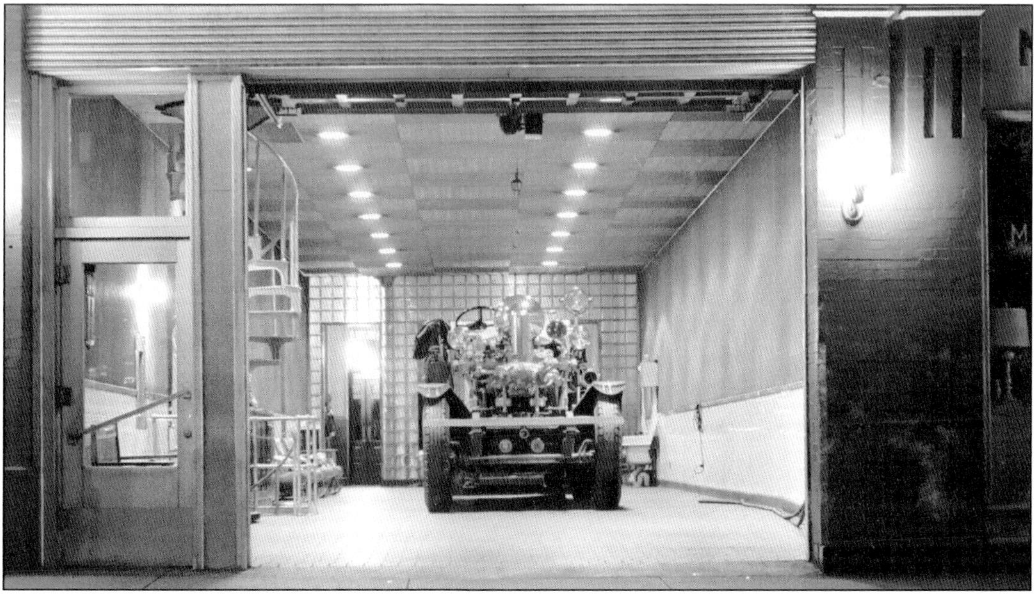

"Dynamite" is seen here in a nighttime photograph of the newly renovated hose company in the late 1940s. The original stairs were replaced with these spiral stairs made by the Duvinage Corporation of Hagerstown. (Courtesy of the First Hagerstown Hose Company.)

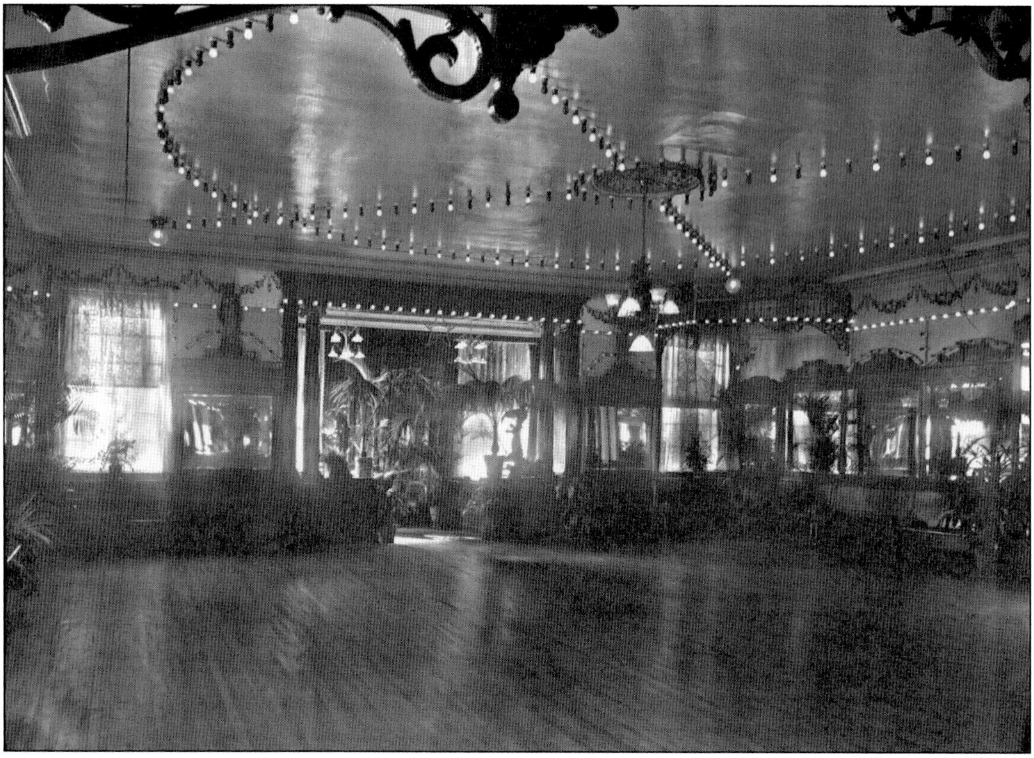

Firehouses built in the late 1800s incorporated the apparatus bay with a large social hall. This is where the term "fire hall" originates. These halls were very ornamental. Many functions, including vaudeville shows, fairs, dances, conventions, and meetings, were held here. In 1923 a photographer's flash bulb ignited nearby decorations. The firefighters had to respond from another incident back to their own firehouse to extinguish the fire. (Courtesy of the First Hagerstown Hose Company.)

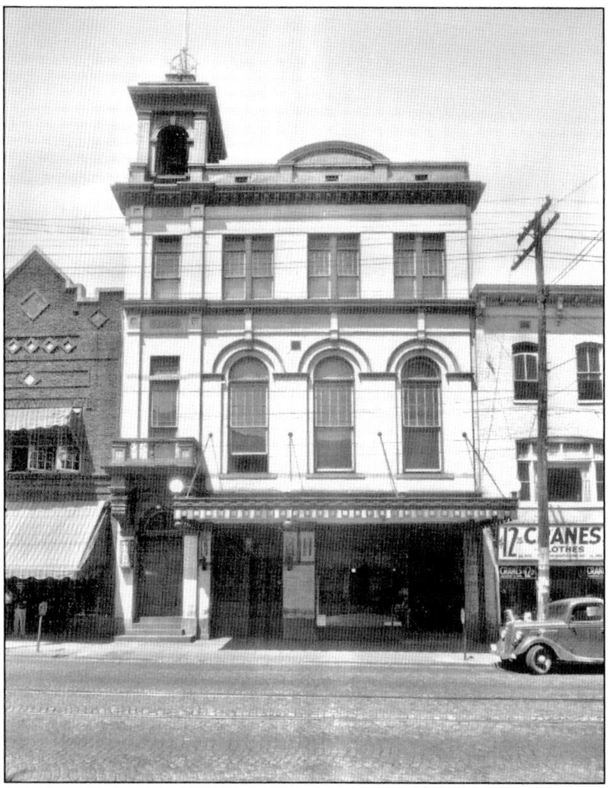

This front view of the hose company was taken in the 1930s. The door on the left is the entrance to the dance hall. During this period, the Marvello Dance Studio occupied the dance hall. How many differences in the exterior of the building now, compared to the 1880s photograph, can you find? (Courtesy of the Washington County Historical Society.)

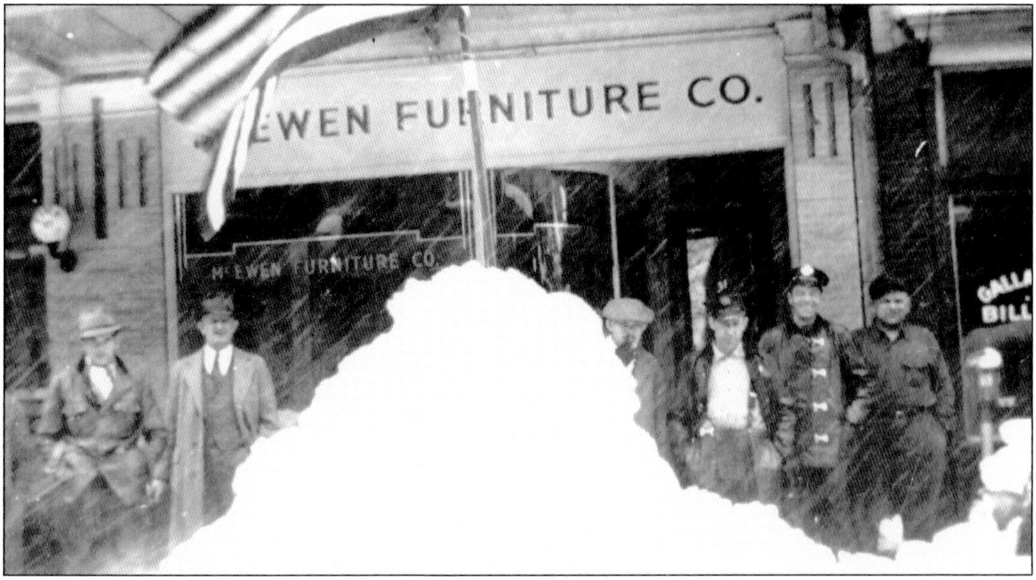

Extreme winter weather conditions hamper efforts to extinguish fires. Firefighters must first remove the snow from in front of the firehouse. Pictured here in a 1942 snowstorm are, from left to right, two unidentified people, Mickey Smith, Jim Bowman, Driver Preston Smith, Driver George Basore, and Pug Schmidt. (Courtesy of the William Dieterich collection.)

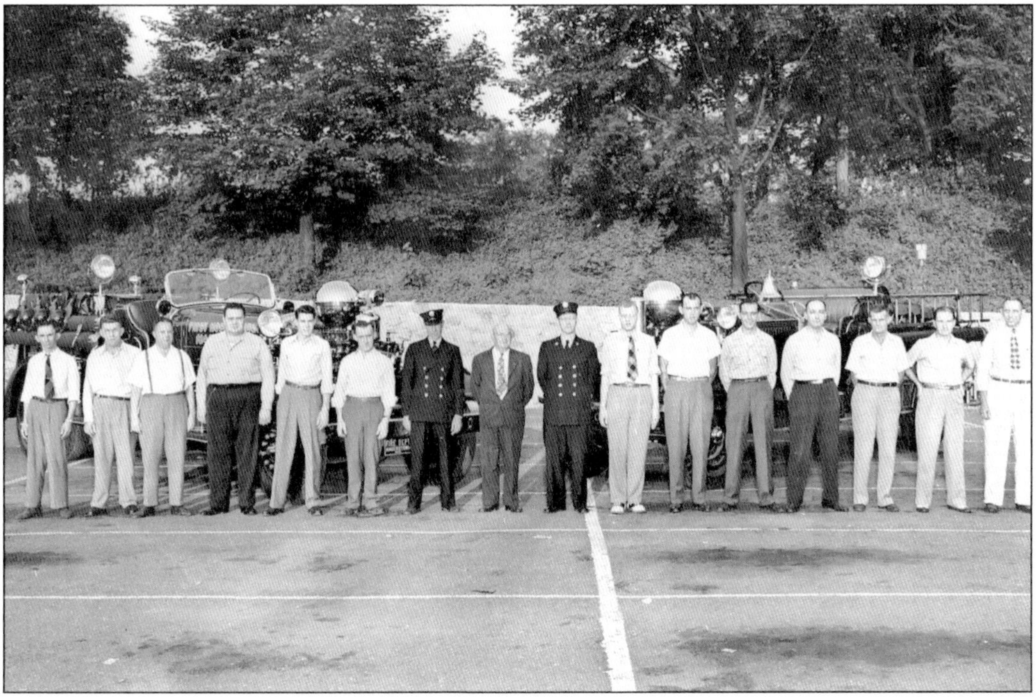

Members of the hose company are standing in front of the fire apparatus in this photograph. From left to right are Willie Overman, ? Downs, Dean Smith, Driver Joe South, John Smith, Ed Beckley, Driver Preston Smith, Jim Mobley, Driver A.K. McGraw, Robert Lewis, ? William Karn, Maurice Domenici, Jim Bowman, Robert Dean, and ? Maisack. (Courtesy of the First Hagerstown Hose Company.)

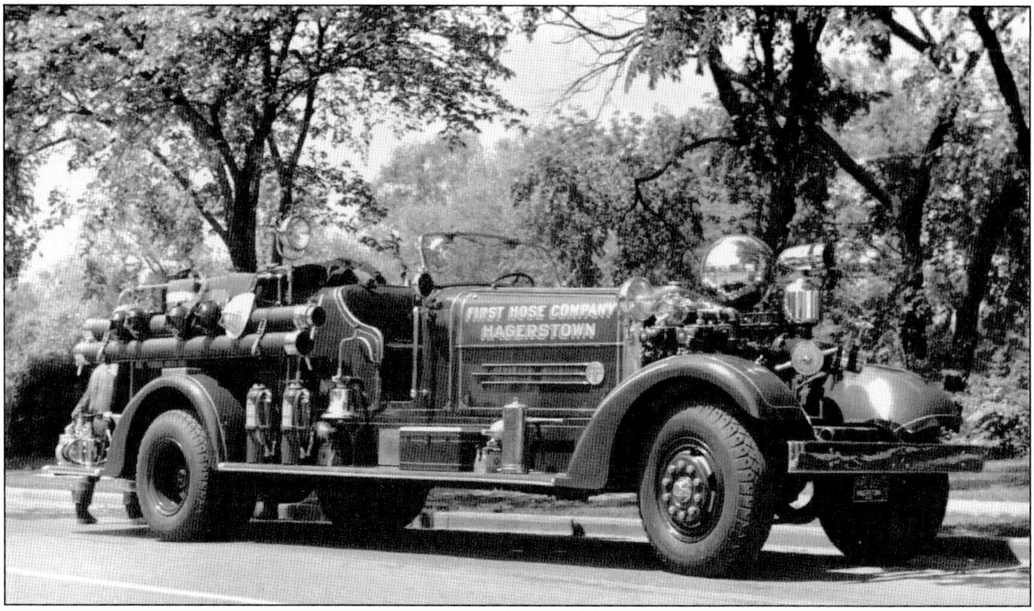

Because of a shortage of metal following World War II, this Ahrens-Fox Model HT (Reg. No. 3459) pumper was the only one completely built by Ahrens-Fox in 1946. It has a 1,000-gallon-per-minute pump and a 100-gallon water tank, weighs 15,100 pounds, and carries the affectionate name "Slugger." It cost $16,474 when the contract was signed on January 16, 1946. (Courtesy of the First Hagerstown Hose Company.)

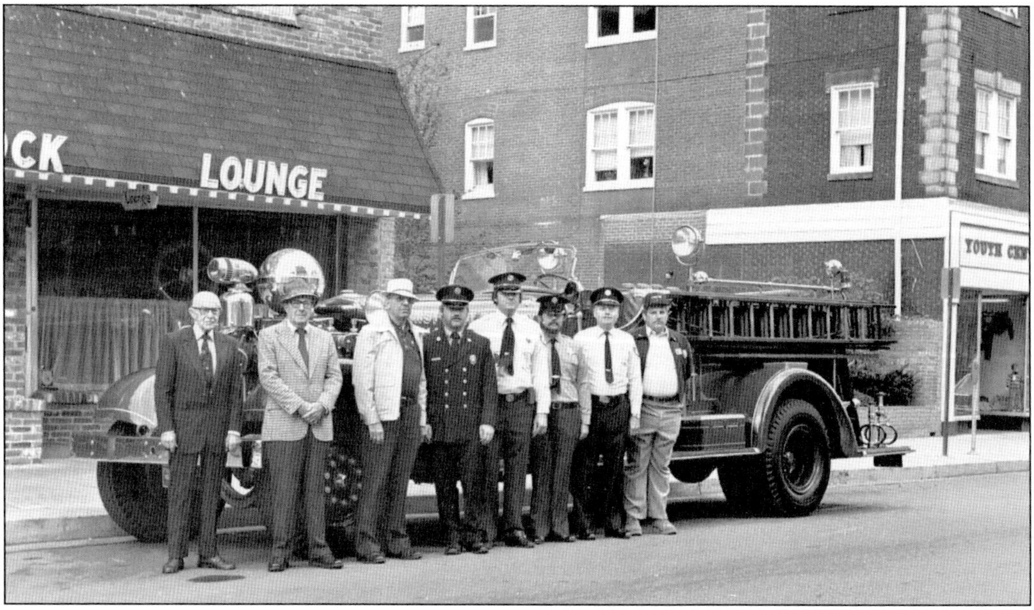

The 1946 Ahrens-Fox was sold to the hose company by the City of Hagerstown for $1. This provided the opportunity to keep a treasured piece of fire apparatus in town. "Slugger" is taken to fire apparatus musters and usually carries the mayor and council through the Alsatia Mummers parade. From left to right are Jesse Berre (who joined the hose company in 1918), Richard Henneberger, Robert Lewis, FAO David Baer, David Stevens, Chris Miller, Blaine Snyder, and Ray Lushbaugh. (Courtesy of the First Hagerstown Hose Company.)

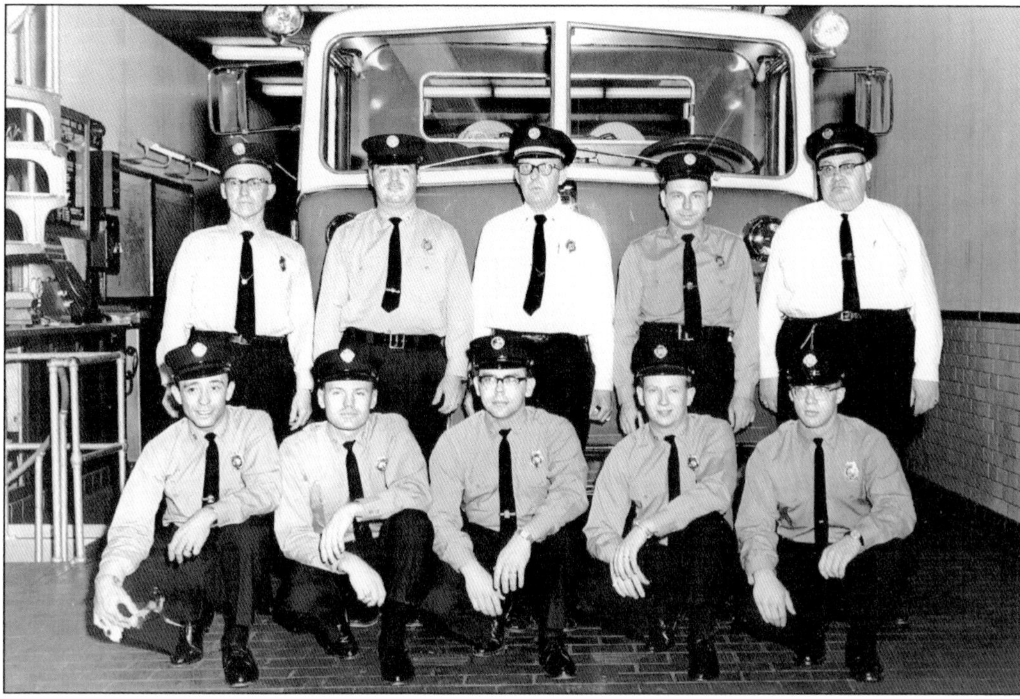

This is a photograph of the operations officers in the 1960s. From left to right are (front row) Ronald Karn, Blaine Snyder, Tom Kaiss, Bob Carder, and Ed Mobley; (back row) Driver A.K. McGraw, Donald Koons, Robert Lewis, Terry Snider, and Driver Joe South. (Courtesy of the First Hagerstown Hose Company.)

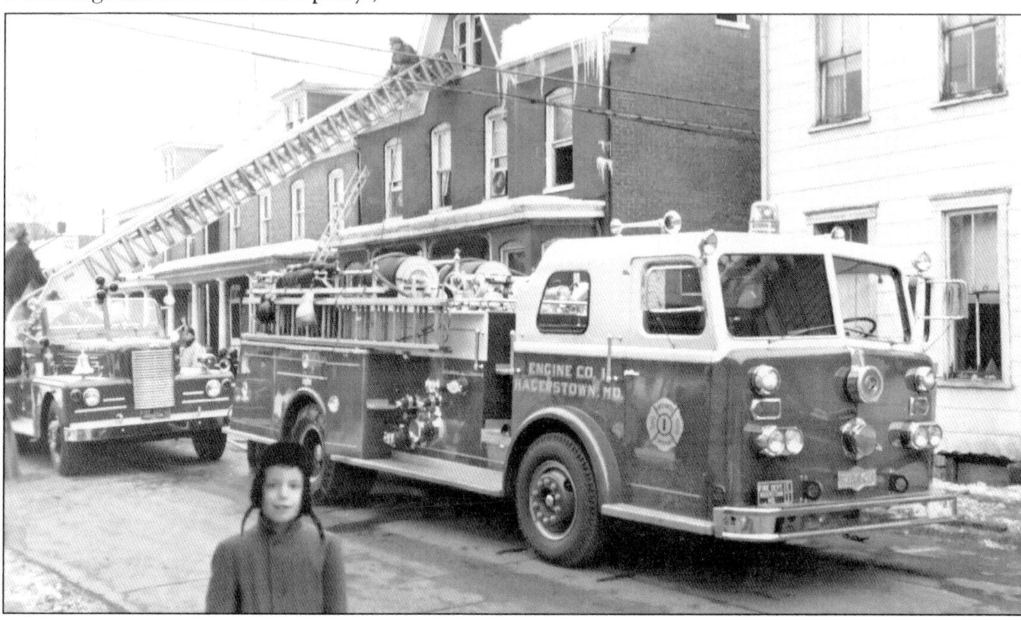

Did this young man grow up and become a firefighter? In the background is the 1963 Pirsch pumper. It has a 1,000-gallon-per-minute pump and a 300-gallon water tank. It is known as "The Boxcar." The photograph was taken at a fire scene on North Locust Street. (Courtesy of the First Hagerstown Hose Company.)

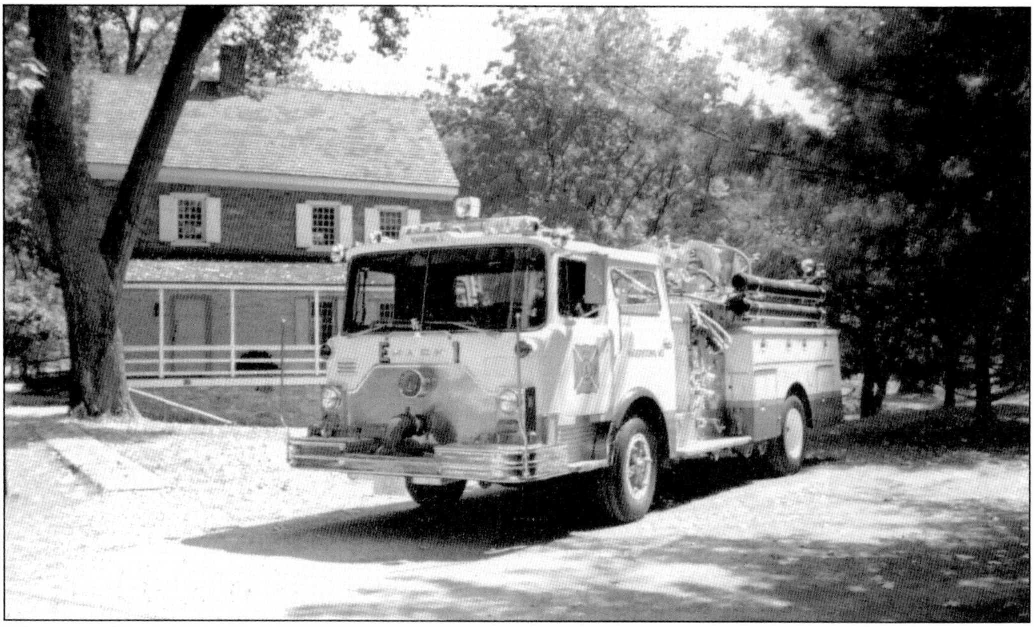

This is a 1982 CF Mack pumper. It is one of three pumpers the city purchased at the same time. It has a 1,000-gallon-per-minute pump and a 500-gallon water tank. In the background is the restored home of Hagerstown's founder, Jonathan Hager. The home was built in 1739. (Courtesy of the author.)

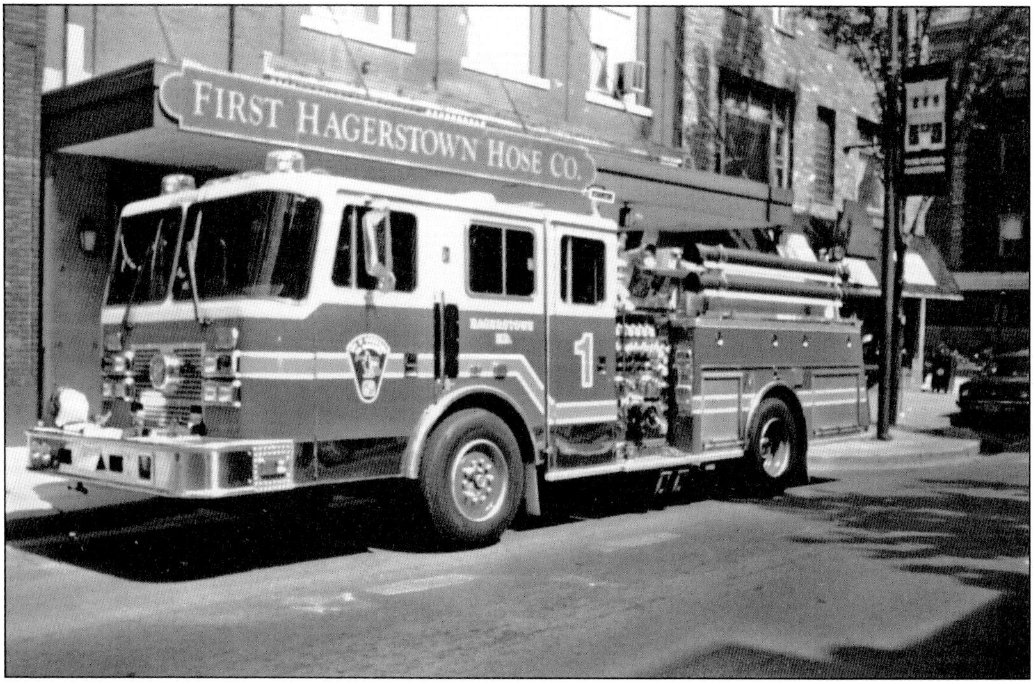

The newest pumper at the hose company is this 1993 KME. It can pump 1,500 gallons per minute and has a 500-gallon water tank. This is the hose company's first fully enclosed cab, which permits all of the firefighters to ride inside. Standing on the back of fire engines while going to fires is now a thing of the past. (Courtesy of Philip Mayhue.)

Two

Antietam Fire Company, Engine 2

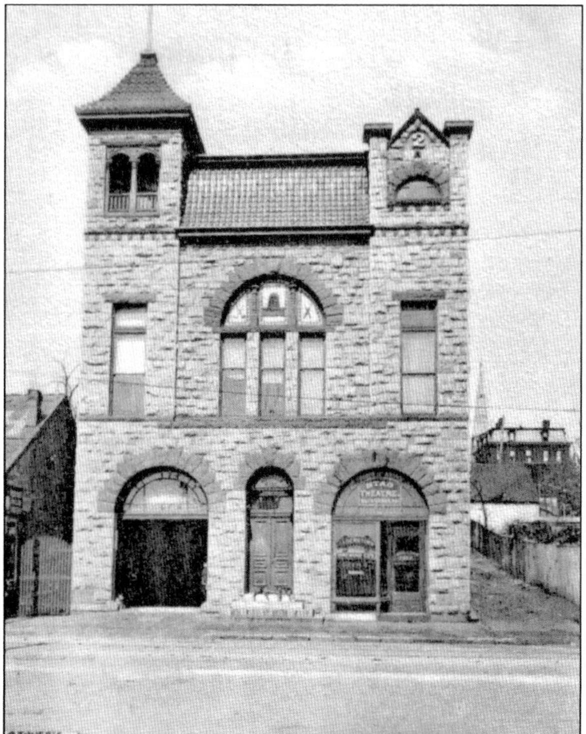

The Antietam Fire Company was organized as early as 1808. Its first known permanent home was in the unit block of Summit Avenue next to the Washington County Court House in about 1835. This building was constructed at 113 Summit Avenue in 1895. Built during the period called the "Golden Age of Fire Station Design" the facade is known to have inspired the design of the War Correspondence Arch in Gapland, Maryland. (Courtesy of the David White collection.)

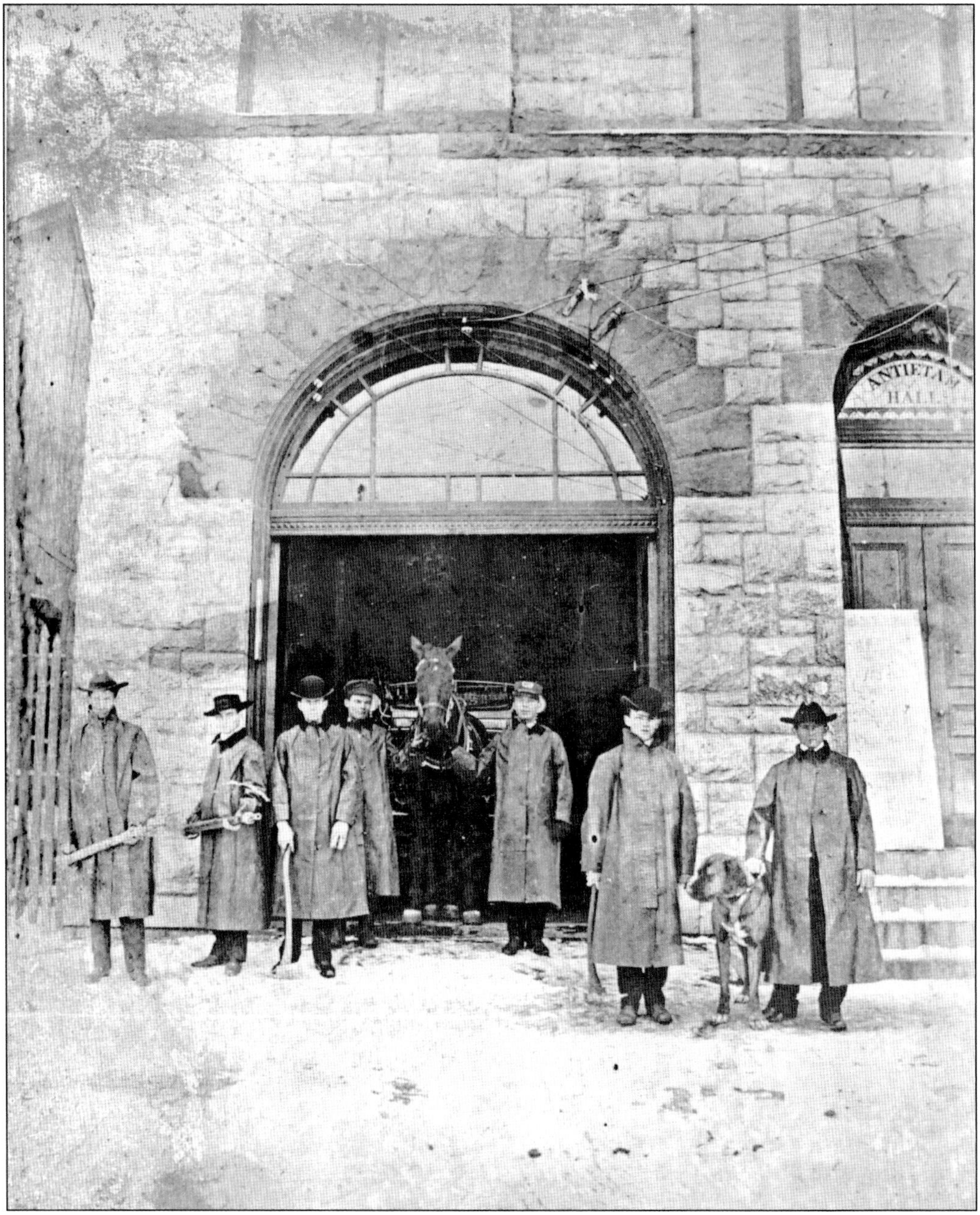

The original name of the Antietam Fire Company was Lafayette. On January 3, 1835, a spirited debate unfolded as to what name should be used in the incorporation document. On the first ballot, "Washington" received 23 votes and "Phoenix" received 18. On January 17 another vote was taken. This time "Antietam" had 32 votes and "Columbian" had 22 votes. The Antietam Fire Company was incorporated the same year. These firefighters are posing in front of the firehouse on January 30, 1904. Note the equipment the firefighters are holding and the size of the dog. (Courtesy of the William Dieterich collection.)

This incredible document is a run report from 1899. The volume of incidents was far less in the 1800s than today. However, when the alarm sounded there was usually a fire. The chief or fire marshal reviewed and signed the reports. John Middlekauff was a well-respected firefighter and assistant fire marshal at the time he signed this report. (Courtesy of the Antietam Fire Company.)

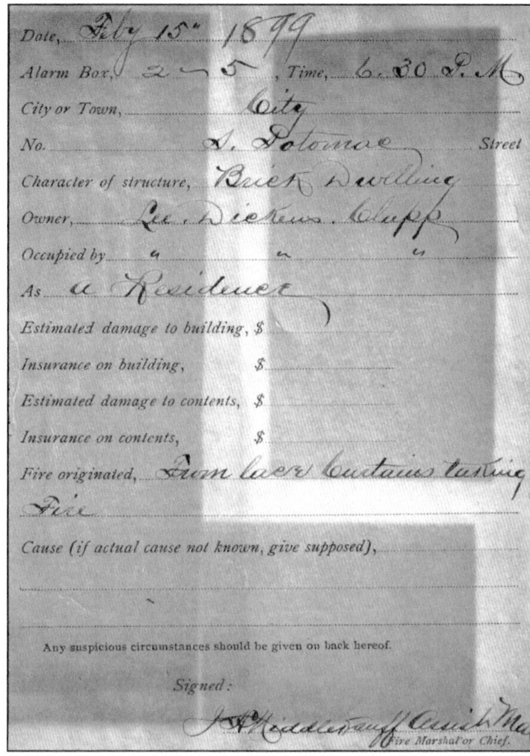

Date, Feby 15" 1899

Alarm Box, 2–5, Time, 6.30 P.M.

City or Town, City

No. S. Potomac Street

Character of structure, Brick Dwelling

Owner, Lee Dickens Clapp

Occupied by " " "

As a Residence

Estimated damage to building, $

Insurance on building, $

Estimated damage to contents, $

Insurance on contents, $

Fire originated, From lace Curtains taking Fire

Cause (if actual cause not known, give supposed),

Any suspicious circumstances should be given on back hereof.

Signed: J. H. Middlekauff Asst. Mar

Fire Marshal or Chief.

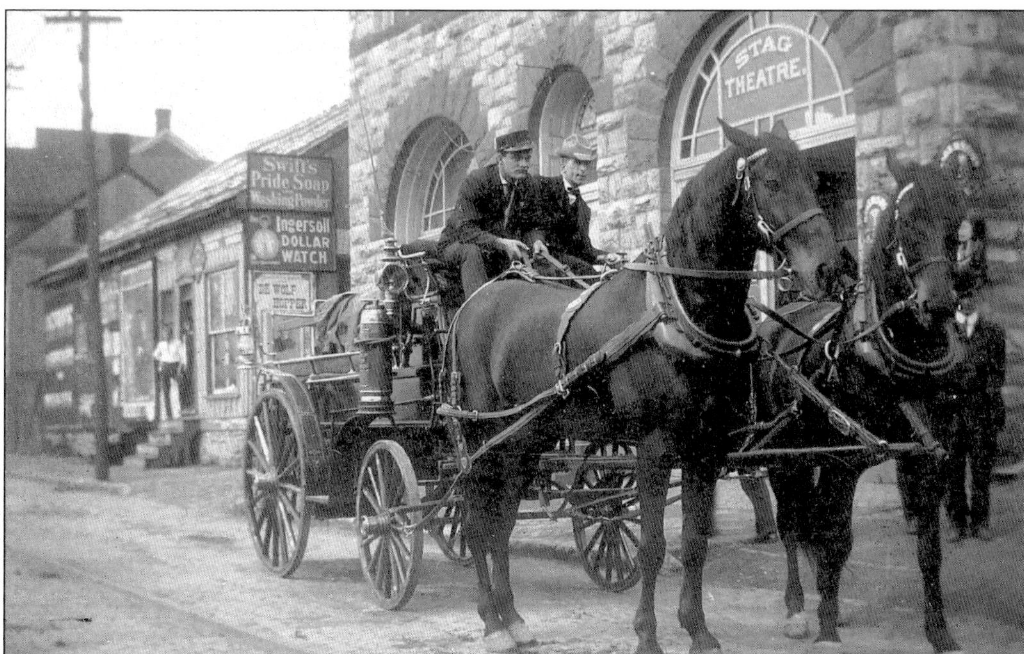

This hose wagon is being pulled by the two-horse team of Colonel and Bill. The driver is a young Charles "Possum" Harrison. Driver Harrison provided 42 years of loyal service to the Antietam Fire Company. Note that the rental property at the time was the Stag Theatre. (Courtesy of the Washington County Historical Society.)

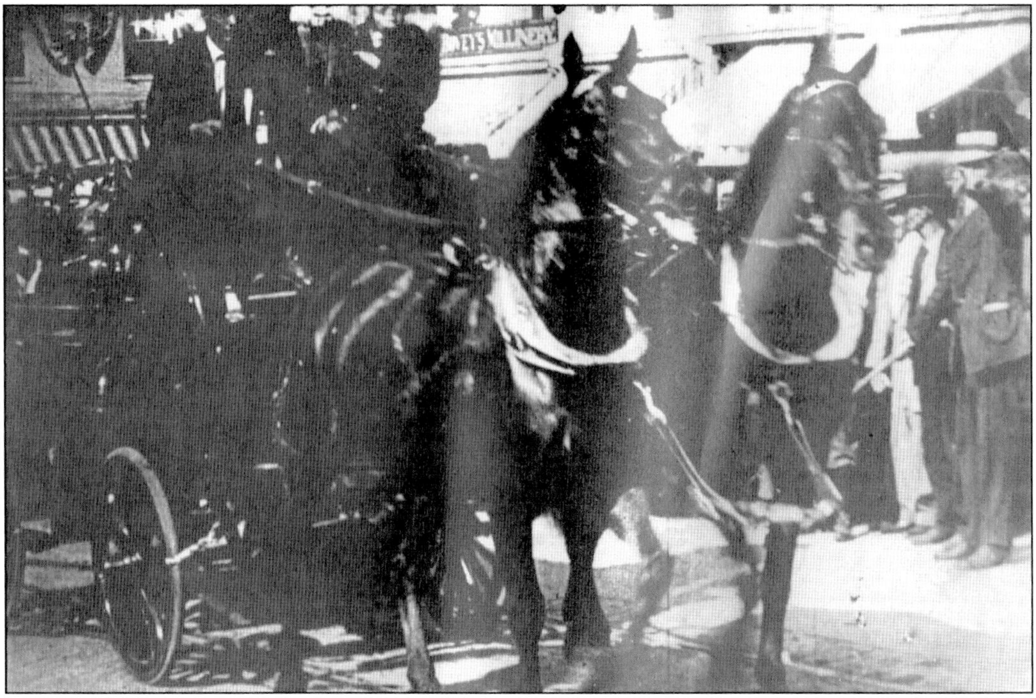

The dual horse team of John and Jim are seen here pulling a hose wagon through a parade. The horses were trained to the sound of the alarm. (Courtesy of the First Hagerstown Hose Company.)

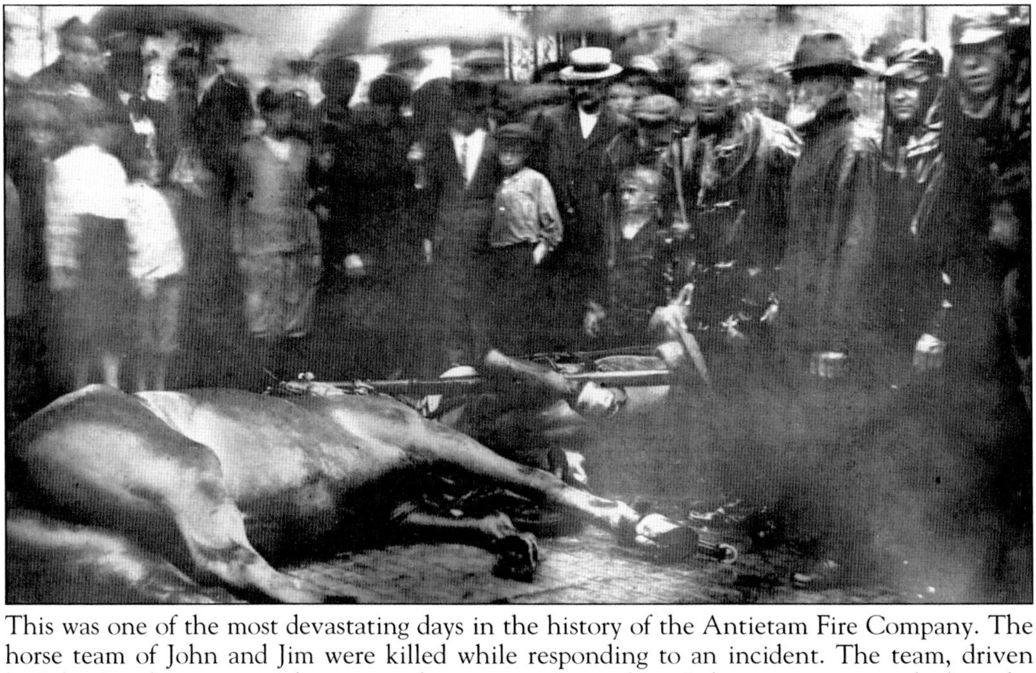

This was one of the most devastating days in the history of the Antietam Fire Company. The horse team of John and Jim were killed while responding to an incident. The team, driven by John Bomberger, turned onto South Potomac Street from Baltimore Street, and when the horses came in contact with a downed power line they fell to the ground instantly. Driver Bomberger fell from the hose wagon as well but survived. The incident took place just after a thunderstorm passed through the area. (Courtesy of the Antietam Fire Company.)

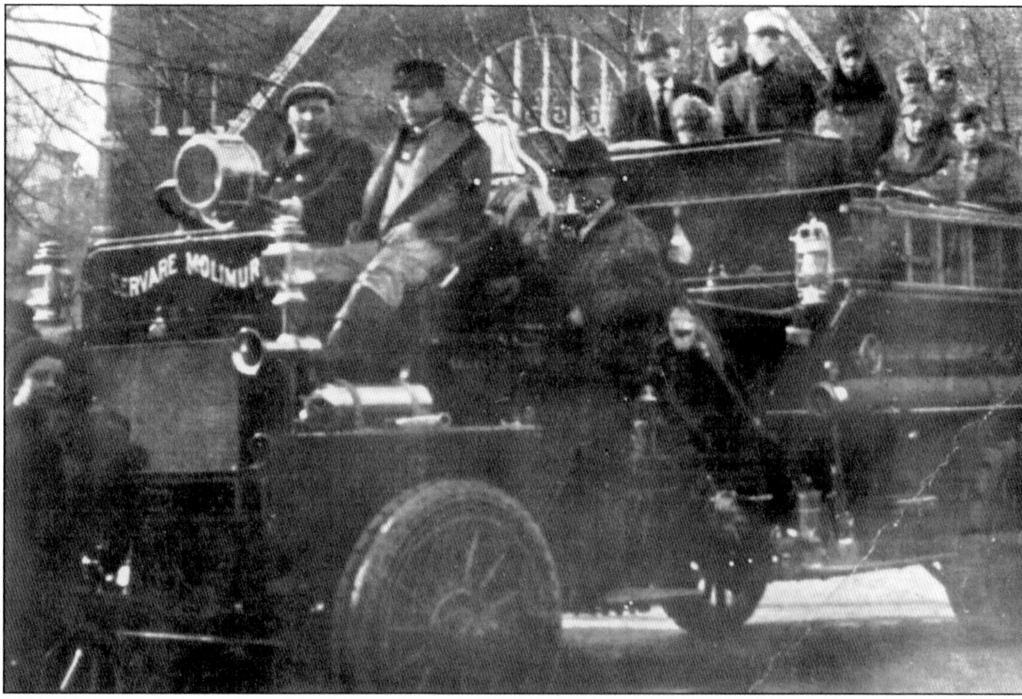

The Antietam Fire Company's first motorized fire apparatus was this 1913 Knox combination pumper. During an acceptance test, the Knox pumped over 600 gallons-per-minute. The Knox had a 90-horsepower motor. (Courtesy of the Antietam Fire Company.)

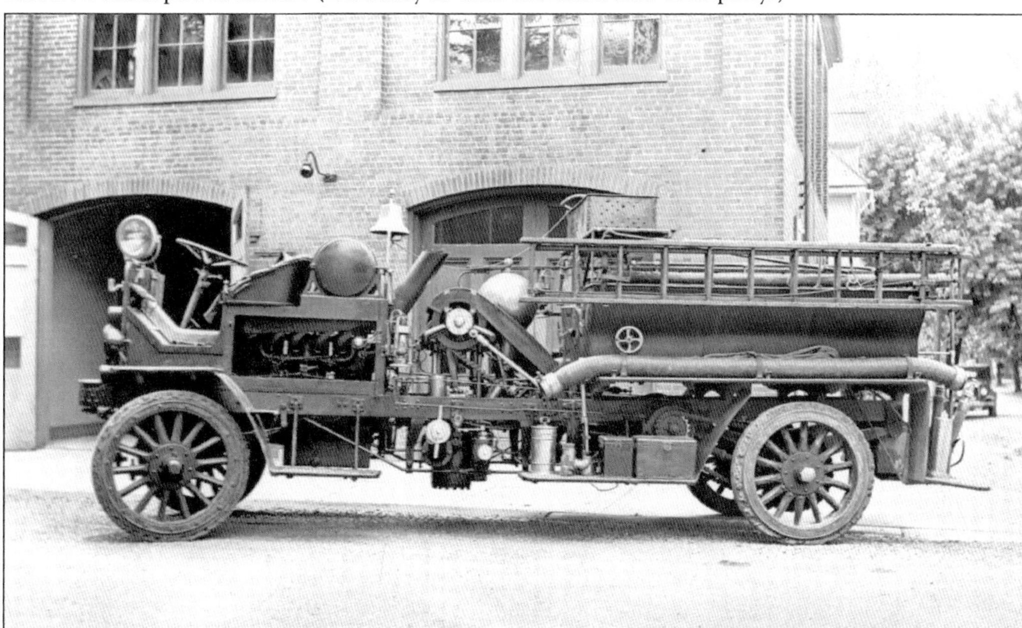

The Knox served Hagerstown until 1927, when it was sold to the Rescue Hose Company No. 1 of Greencastle, Pennsylvania. In 1938 the Knox was sold to a local junk dealer. This photograph was taken in front of the Rescue Hose Company's firehouse located at the intersection of Washington and Madison Streets. (Courtesy of the Ray Mowen collection.)

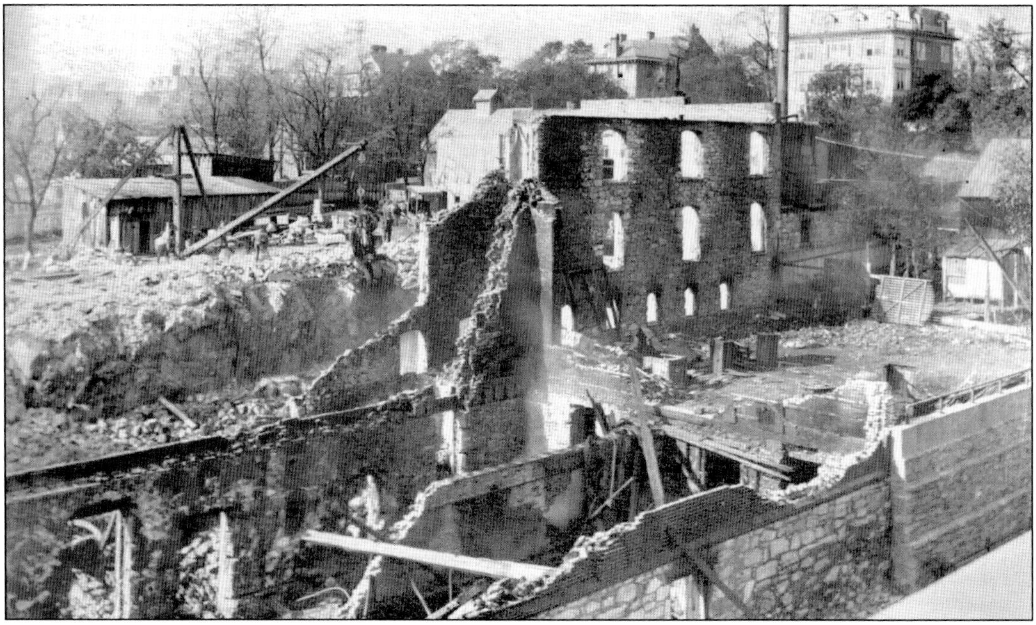

One of the most disastrous days in Hagerstown history occurred on October 28, 1914. A fire started in the Antietam Garage, located on Summit Avenue next to the Antietam Fire Company. It is believed that a spark ignited gasoline fumes and caused an explosion. The fire quickly spread next door, consuming the Antietam Fire Company. Note the auto being pulled from the ruins of the Antietam Garage. (Courtesy of the First Hagerstown Hose Company.)

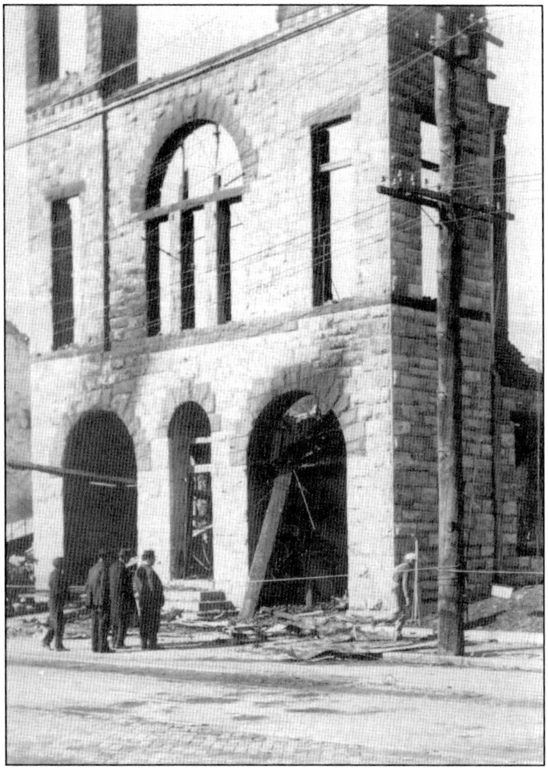

The only part of the structure that is recognizable in this 1914 photograph is the front facade. Six years later another fire would strike the building. The fire company rebuilt using the original facade. Evidence of both fires is still visible today in the attic area. (Courtesy of the First Hagerstown Fire Company.)

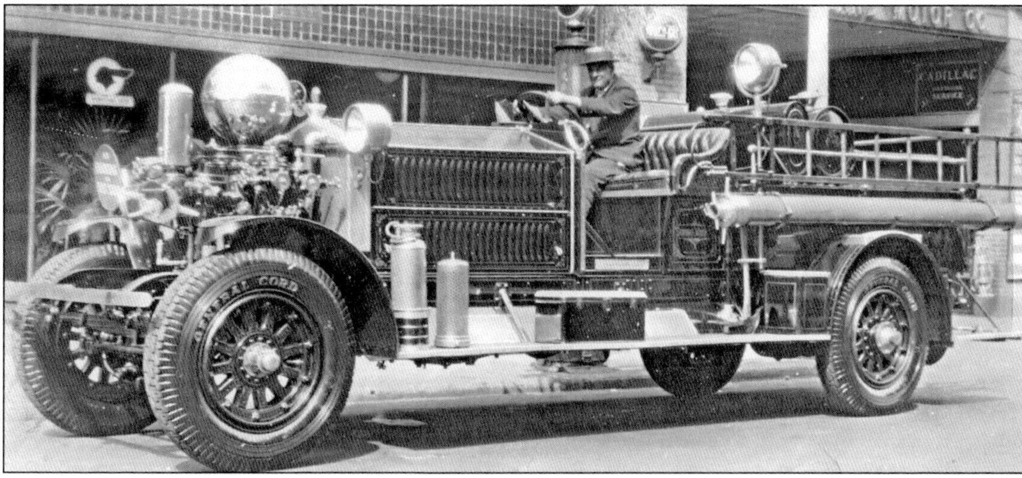

This Ahrens-Fox Model N-S-4 (Reg. No. 1609) pumper was shipped on March 21, 1924. It has a 1,000-gallon-per-minute piston pump and a 60-gallon water tank. It is affectionately known as "The Queen Mary." The driver wearing a straw hat is Charles "Possum" Harrison. (Courtesy of the Rick Delinger collection.)

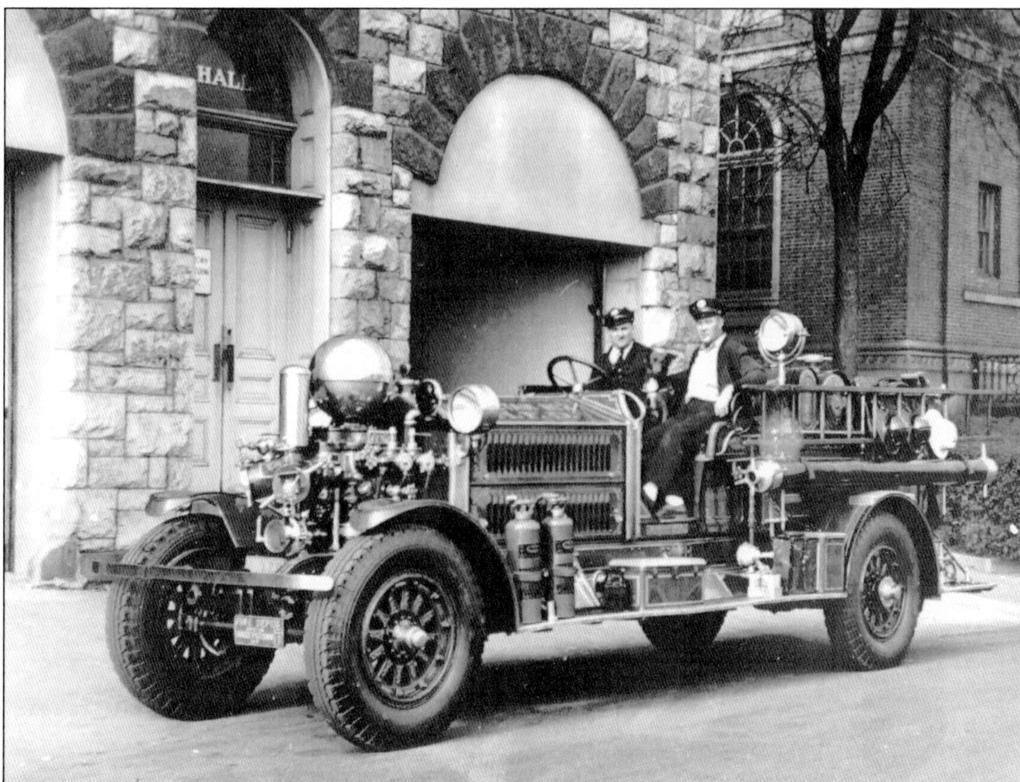

This is another view of the 1924 Ahrens-Fox in front of the firehouse. The drivers are Ralph Foltz and Kenny Myers. Pumpers that had a front-mount pump were difficult to maneuver through narrow city streets. The pump that was positioned well ahead of the cab made it hard to see past the bumper. The pumpers were not equipped with power steering. (Courtesy of the First Hagerstown Hose Company.)

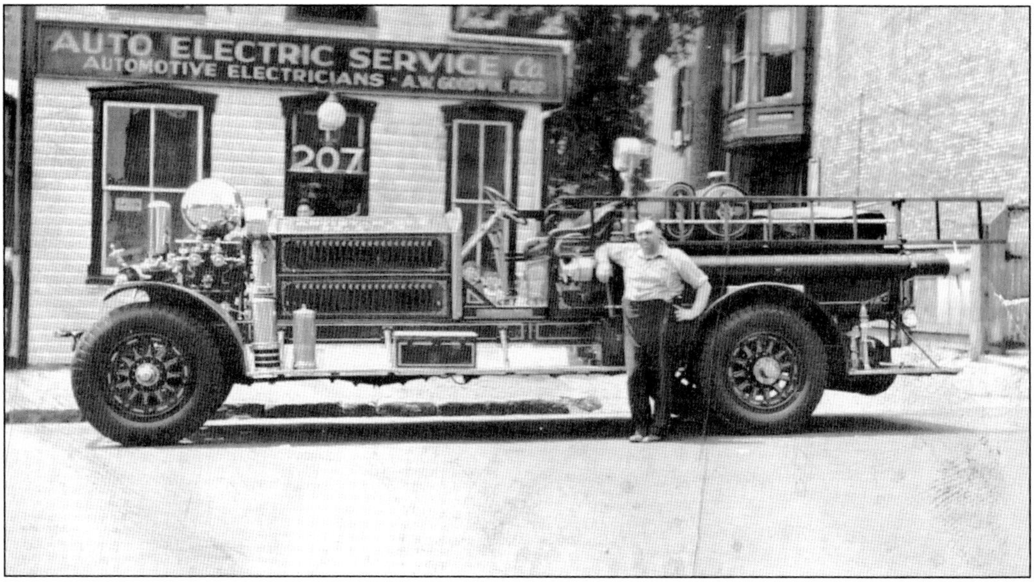

In this photograph John Bomberger poses with the 1924 Ahrens-Fox. Driver Bomberger had the unique opportunity of driving the horses as well as the motorized apparatus. Driver Bomberger worked as a pipe maker at Moller Organ Works before becoming a firefighter. He was born on December 1, 1888, and he joined the fire department in 1906. (Courtesy of the Antietam Fire Company.)

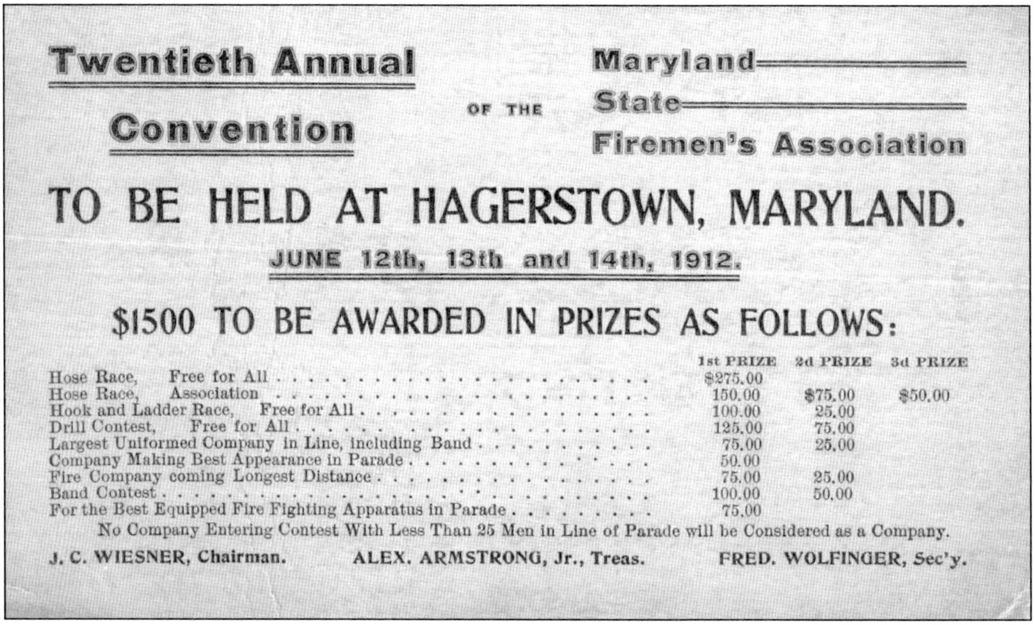

Twentieth Annual Convention OF THE **Maryland State Firemen's Association**

TO BE HELD AT HAGERSTOWN, MARYLAND.

JUNE 12th, 13th and 14th, 1912.

$1500 TO BE AWARDED IN PRIZES AS FOLLOWS:

	1st PRIZE	2d PRIZE	3d PRIZE
Hose Race, Free for All	$275.00		
Hose Race, Association	150.00	$75.00	$50.00
Hook and Ladder Race, Free for All	100.00	25.00	
Drill Contest, Free for All	125.00	75.00	
Largest Uniformed Company in Line, including Band	75.00	25.00	
Company Making Best Appearance in Parade	50.00		
Fire Company coming Longest Distance	75.00	25.00	
Band Contest	100.00	50.00	
For the Best Equipped Fire Fighting Apparatus in Parade	75.00		

No Company Entering Contest With Less Than 25 Men in Line of Parade will be Considered as a Company.

J. C. WIESNER, Chairman. ALEX. ARMSTRONG, Jr., Treas. FRED. WOLFINGER, Sec'y.

Not only was the reel contest a test of a fire company's abilities, it also paid out good prize money. The $250 prize for first place in the free-for-all hose race would be considered a large sum of money even by today's standards. (Courtesy of the Antietam Fire Company.)

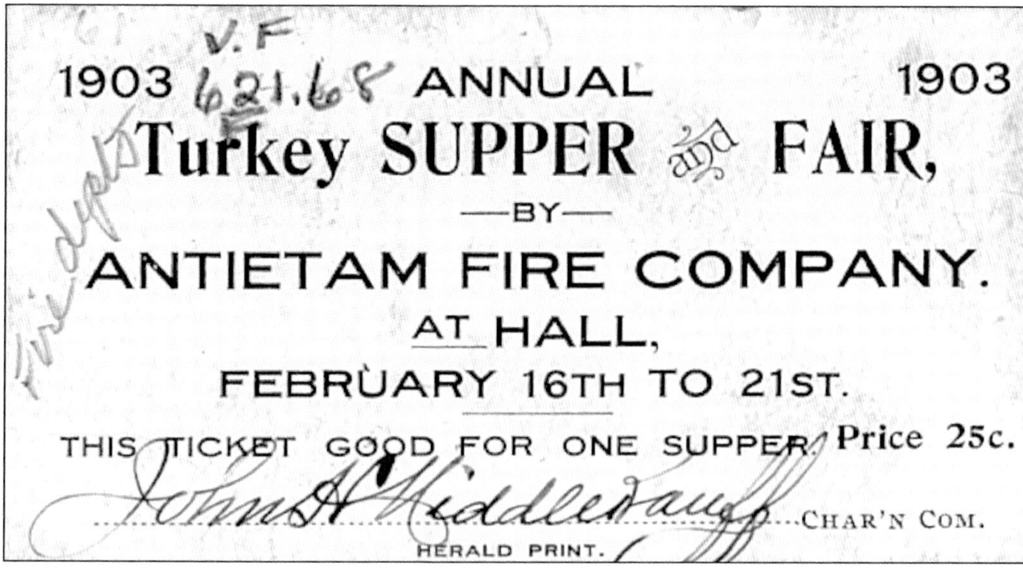

1903 ANNUAL 1903

Turkey SUPPER and FAIR,

—BY—

ANTIETAM FIRE COMPANY.

AT HALL,

FEBRUARY 16TH TO 21ST.

THIS TICKET GOOD FOR ONE SUPPER. Price 25c.

..........CHAR'N COM.

HERALD PRINT.

Meals and fairs were not only a good community event, they also helped raise money for the operations of the fire company. One would be hard pressed to find a turkey supper and fair today for just 25¢. (Courtesy of the Washington County Historical Society.)

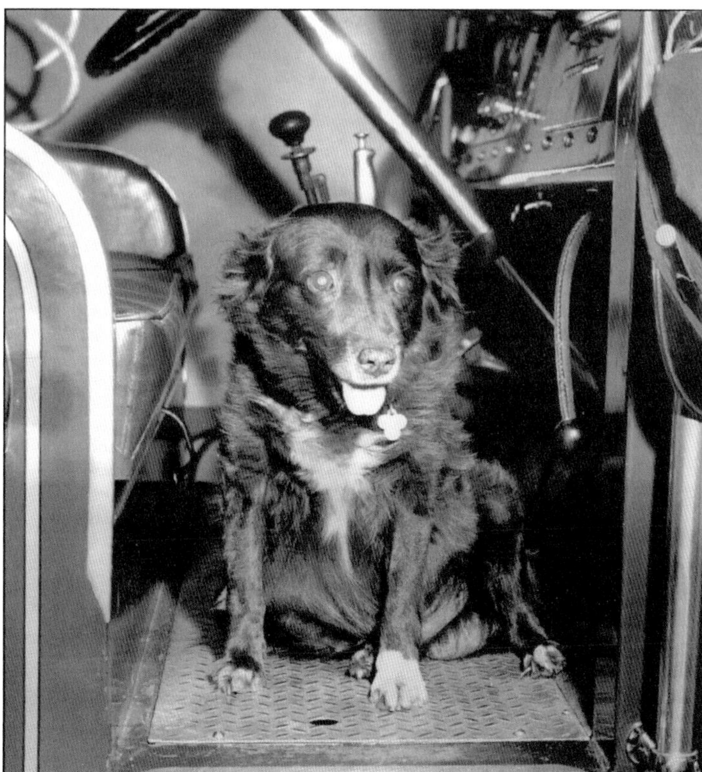

A dalmatian is not the only breed of dog that can be a firehouse mascot. Pictured in this photograph is Moose, the Antietam Fire Company mascot. Moose was a stray that wandered into the firehouse during a thunderstorm. The firefighters felt sorry for the wet, pitiful looking dog, so they made Moose part of the firehouse family. When the alarm was sounded, Moose would watch the driver closely. If the driver reached for his hat and coat the dog would jump on the fire engine and go along to the fire. (Courtesy of the Antietam Fire Company.)

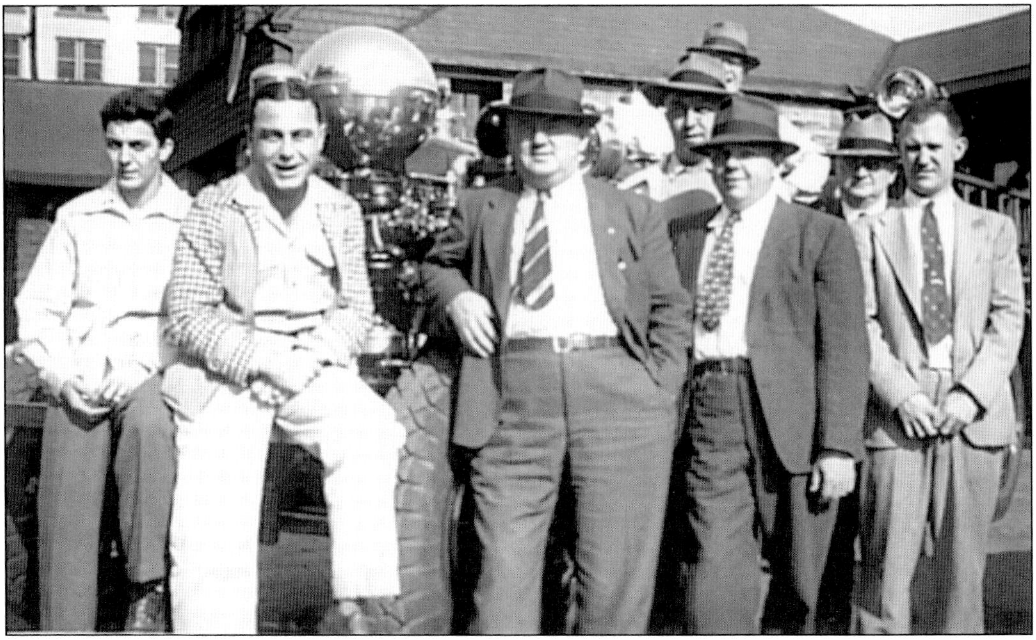

Some of the prominent members of the Antietam Fire Company in 1942, from left to right, are William Dieterich, Rodney Wolfe, George Harper, Leroy Metz, unidentified, Doc Stouffer, and Joe McElroy. The man standing in the back is unidentified. (Courtesy of the Antietam Fire Company.)

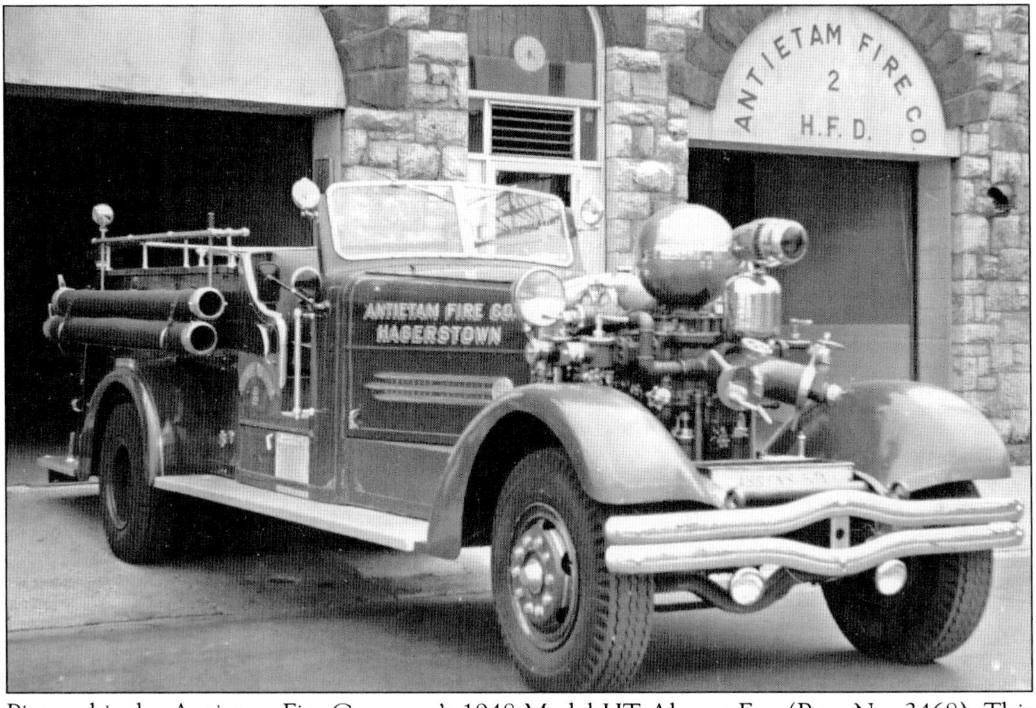

Pictured is the Antietam Fire Company's 1948 Model HT Ahrens-Fox (Reg. No. 3468). This pumper was shipped to the city on November 29, 1948. It has a 1,000-gallon-per-minute pump and a 150-gallon water tank. (Courtesy of the Antietam Fire Company.)

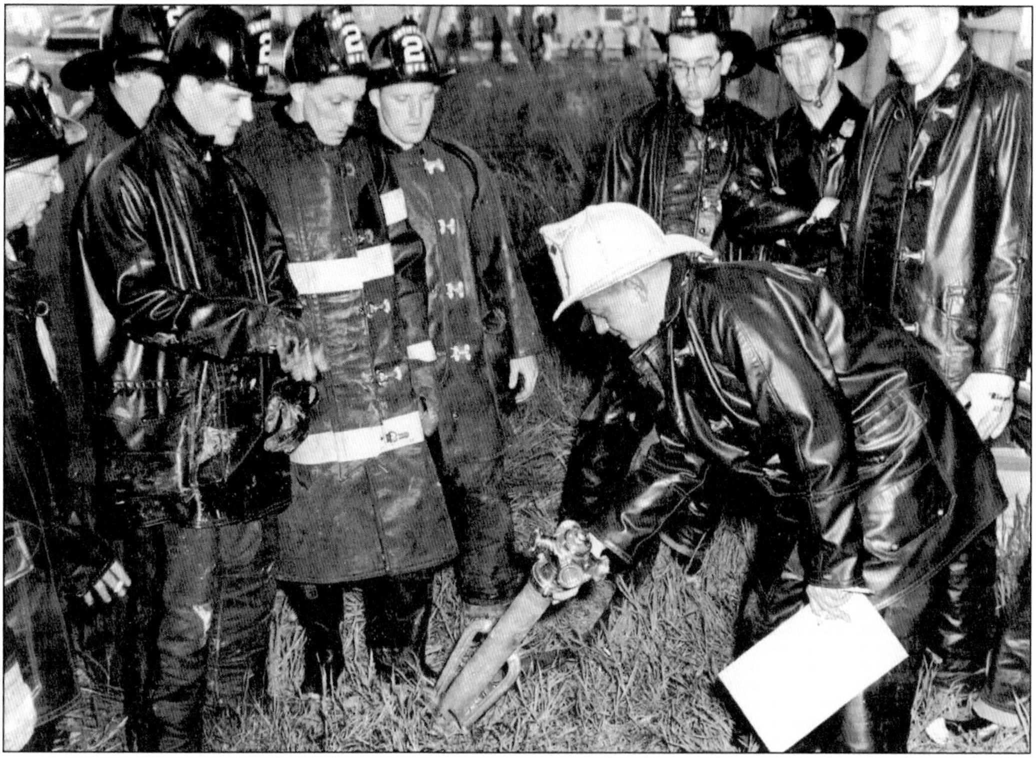

Before the city paid career officers in 1948, each fire company elected a chief. One of the chiefs was then elected as overall chief of the fire department. In addition to managing fires, other responsibilities included training the firefighters. In this photograph Antietam Chief Jack Zinkand is instructing a group of firefighters on the use of a cellar nozzle. (Courtesy of the Antietam Fire Company.)

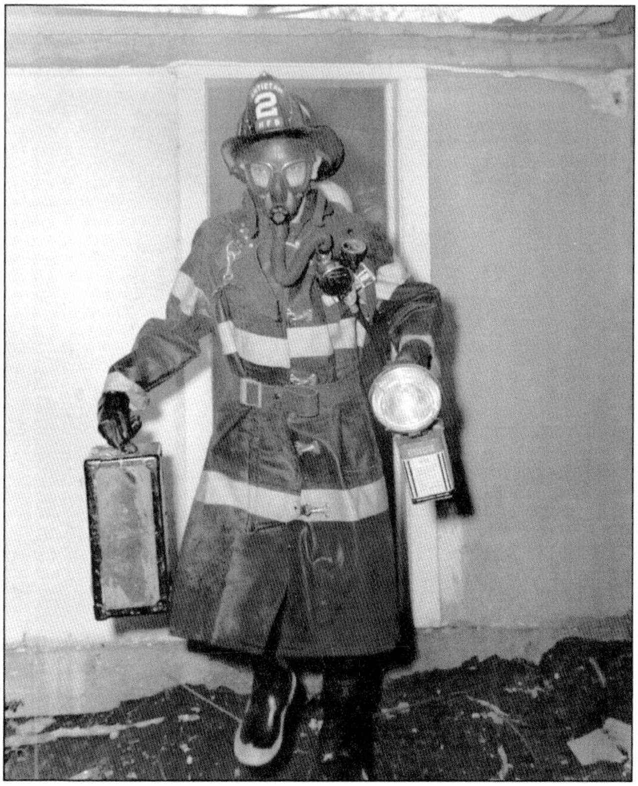

In the early days of firefighting, the men were referred to as "leather lungers" because of the lack of respiratory protection. In this photograph is an Antietam firefighter wearing an early-model Scott-brand, self-contained breathing apparatus (SCBA). (Courtesy of the Antietam Fire Company.)

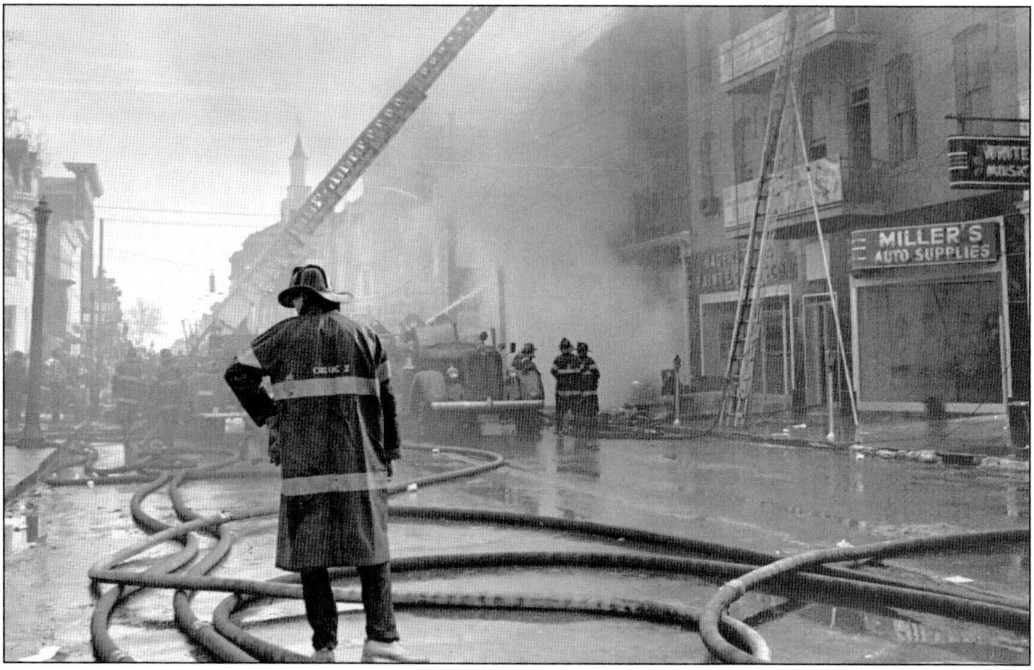

A lone, weary firefighter ponders the situation at the Myers/Berkson furniture store fire on January 9, 1963. Note the 50-foot pole/bangor extension ladder leaning against the Hagerstown Paint and Glass building. (Courtesy of the First Hagerstown Hose Company.)

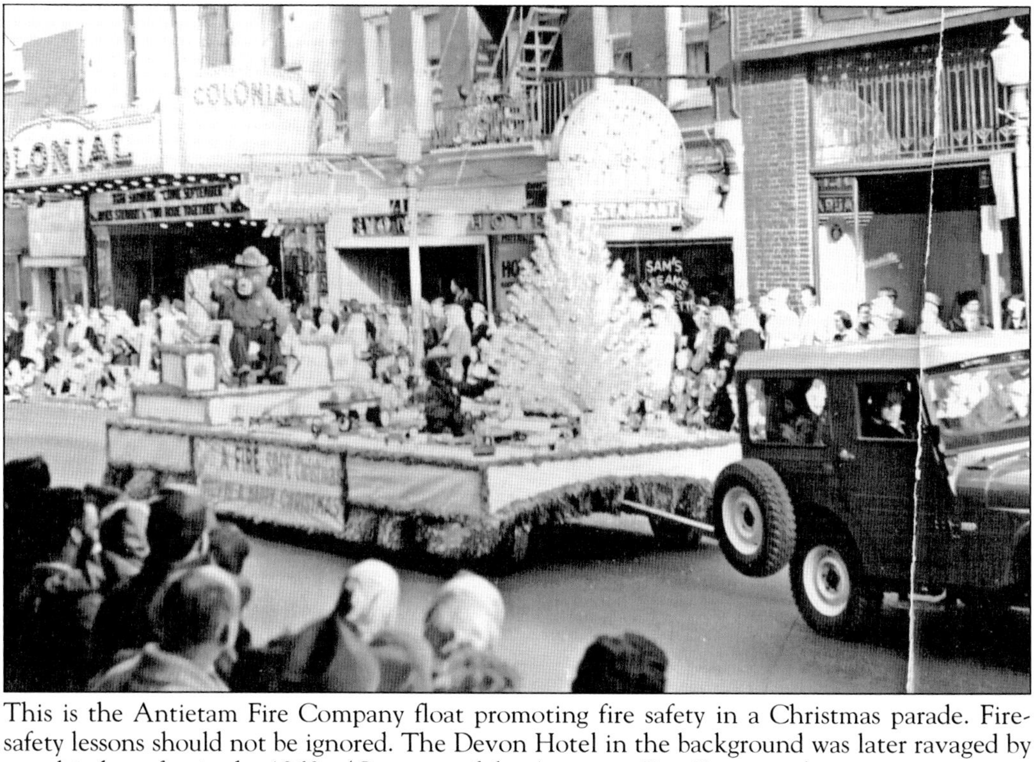

This is the Antietam Fire Company float promoting fire safety in a Christmas parade. Fire-safety lessons should not be ignored. The Devon Hotel in the background was later ravaged by a multi-alarm fire in the 1960s. (Courtesy of the Antietam Fire Company.)

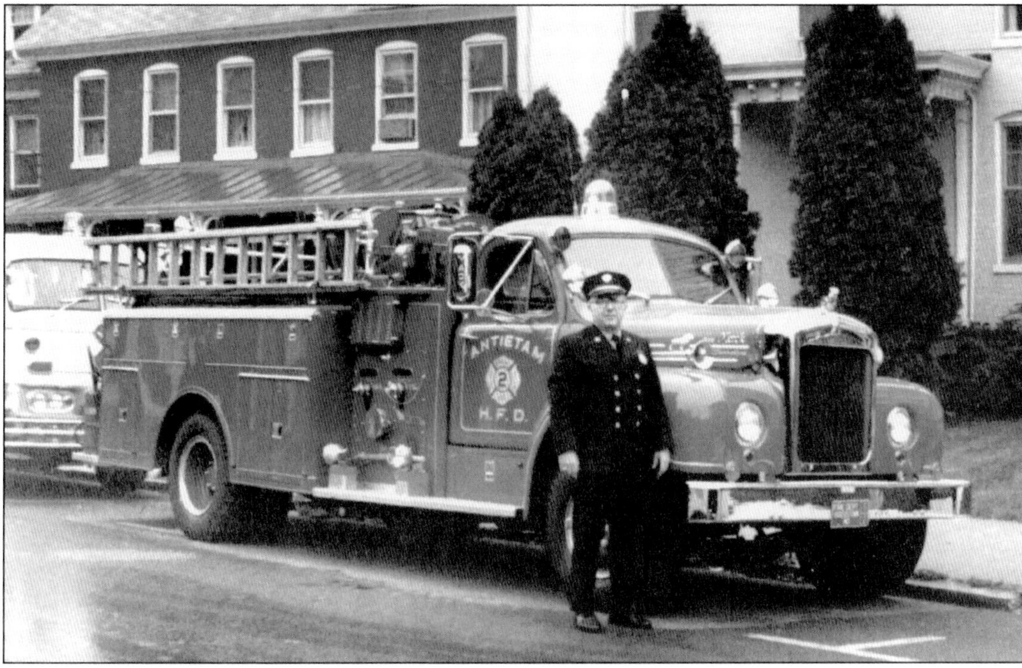

Pictured here with the 1961 "B" model Mack is FAO Fredrick White. FAO White began his career as a driver in 1962. He also served as the treasurer of Local 1605 and was a charter member of the Washington County Emergency Air Unit. FAO White died suddenly while on duty on March 16, 1988. (Courtesy of the Antietam Fire Company.)

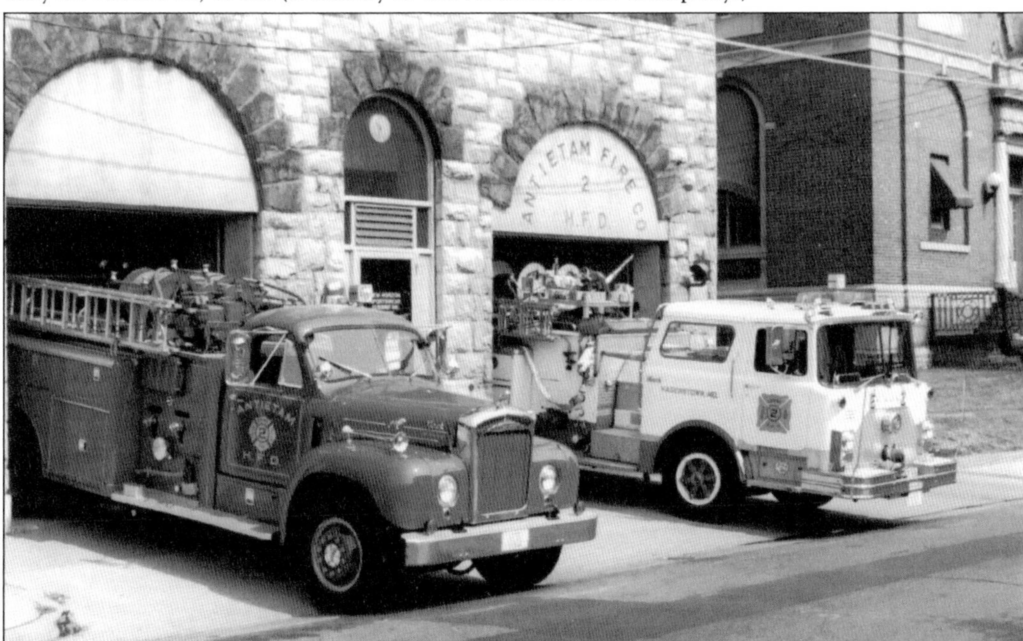

Here is an Antietam Fire Company tandem of apparatus. On the left is a 1961 "B95" model, 1,000-gallon-per-minute pumper. On the right is a 1982 CF Mack, 1,000-gallon-per-minute pumper. (Courtesy of the John Floyd collection.)

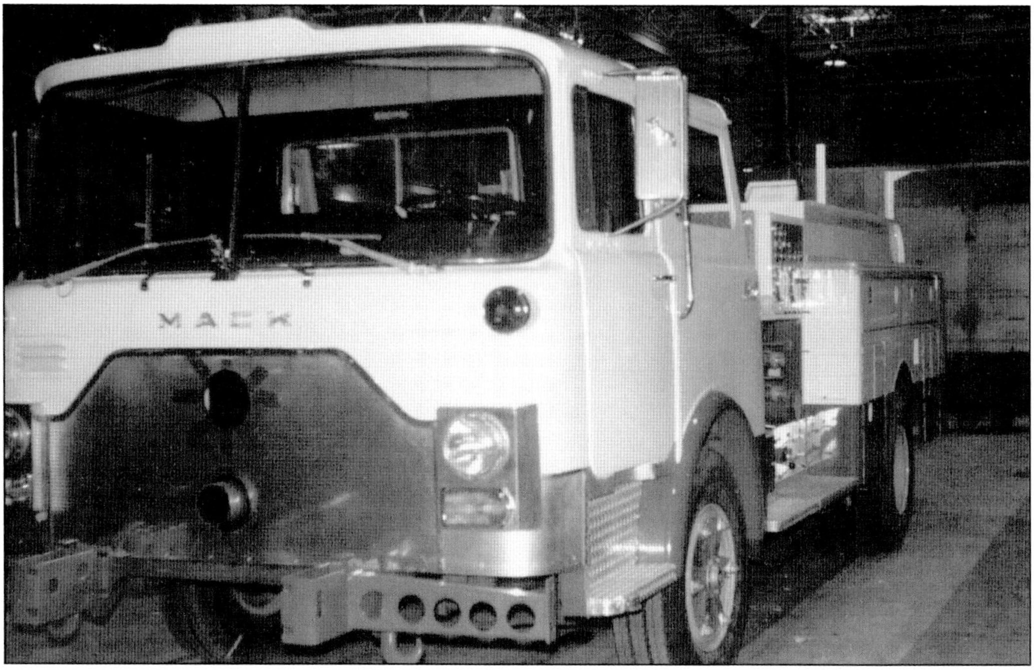

Most fire equipment is customized. A set of detailed specifications are sent to manufacturers for review and a bid. Fire department representatives visit the factory at least twice during the assembly. This is one of three Macks purchased at the same time in 1982. (Courtesy of the Richard Smith collection.)

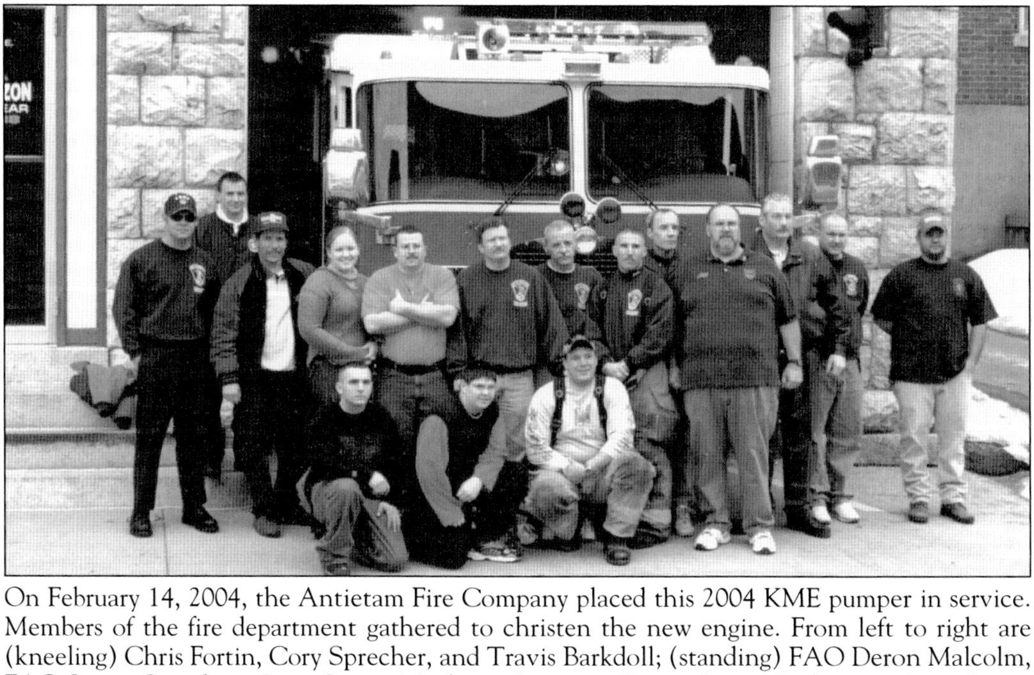

On February 14, 2004, the Antietam Fire Company placed this 2004 KME pumper in service. Members of the fire department gathered to christen the new engine. From left to right are (kneeling) Chris Fortin, Cory Sprecher, and Travis Barkdoll; (standing) FAO Deron Malcolm, FAO James Sprecher, Capt. Justin Mayhue, Courtney Drury, James Forberger, Capt. James Sprecher, Ed Bryan, Toby Higgins, Craig Sheppard, Pres. Jeff Snyder, Harold Semler, Jeff Snyder, and Tom Morris. (Courtesy of the Diana Mayhue collection.)

Three

Independent Junior Fire Company, Engine 3

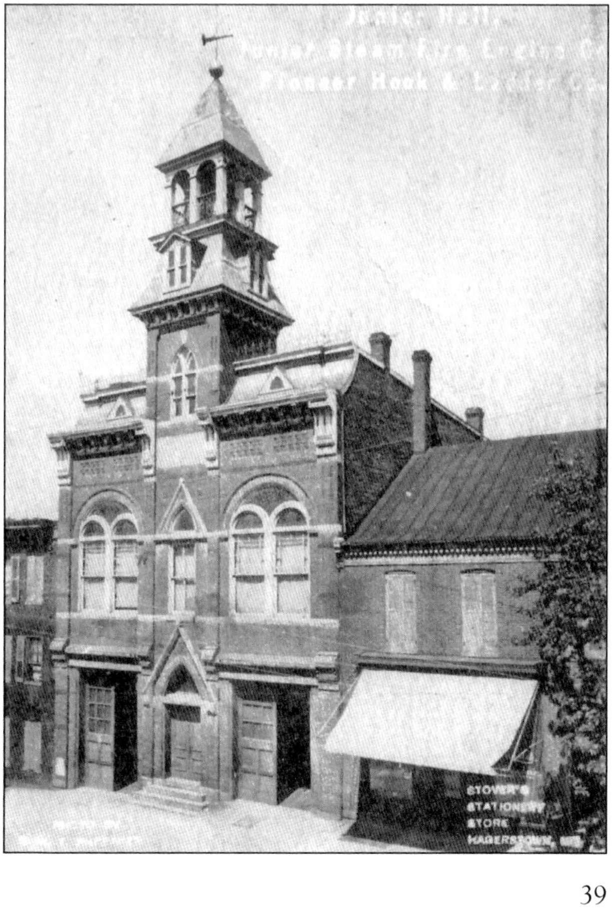

The Junior Fire Company was formed in 1842 by the young members of the Franklin Union Fire Company. The Franklin's engine was also acquired by the young men. The Juniors were incorporated on February 6, 1843. The company motto was "Here We Are." (Courtesy of the Independent Junior Fire Company.)

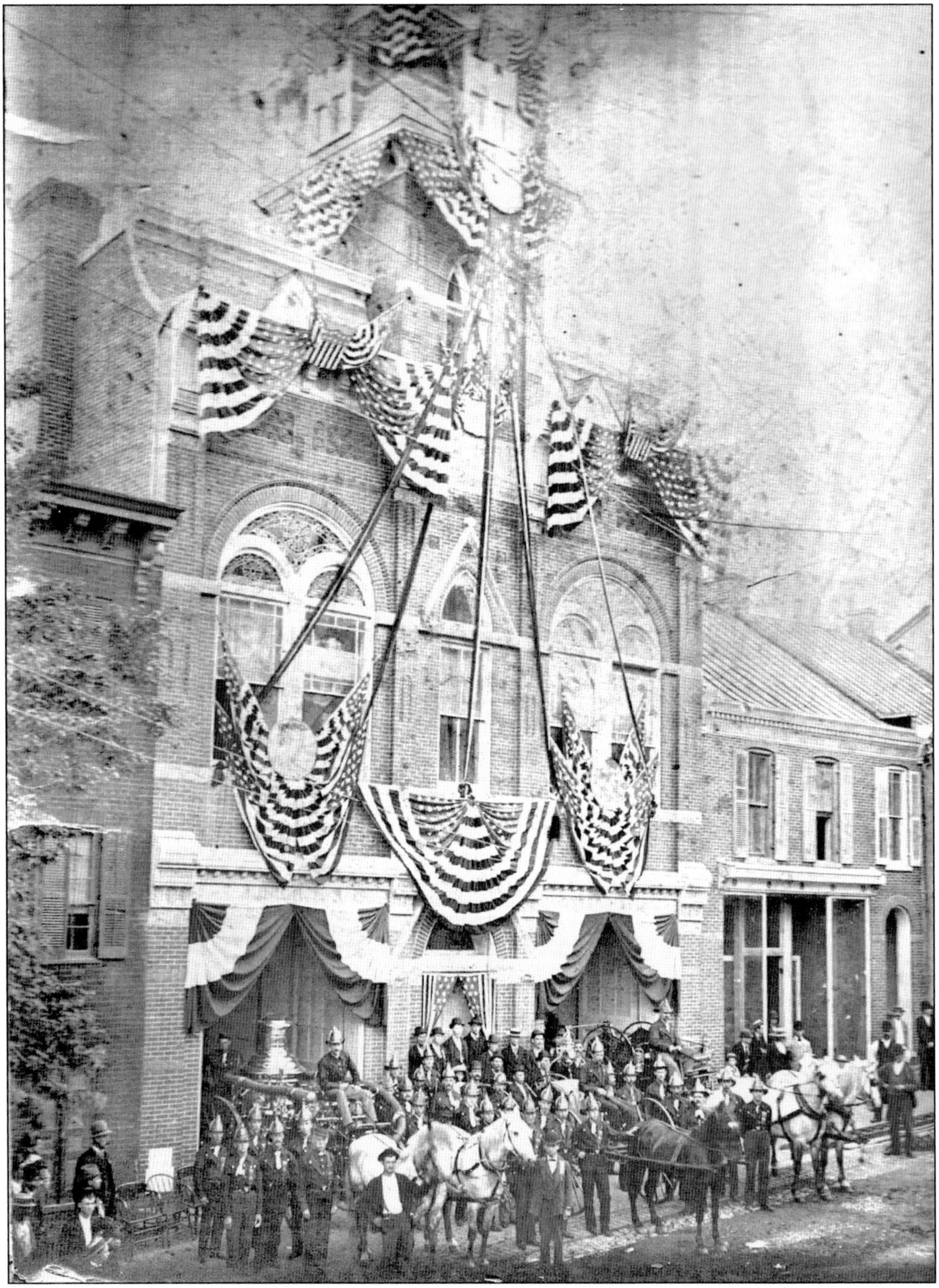

The cornerstone for the original firehouse was laid on May 30, 1852. The building was remodeled in 1892. The members were proud of their firehouse and are seen here during a break in the celebration. The fire apparatus in the photograph includes a steamer and two hose carriages. The Pioneer Hook & Ladder Company is inscribed above the right-hand door. Note that the 100-block of North Potomac Street is dirt. (Courtesy of the Jerry Cunningham collection.)

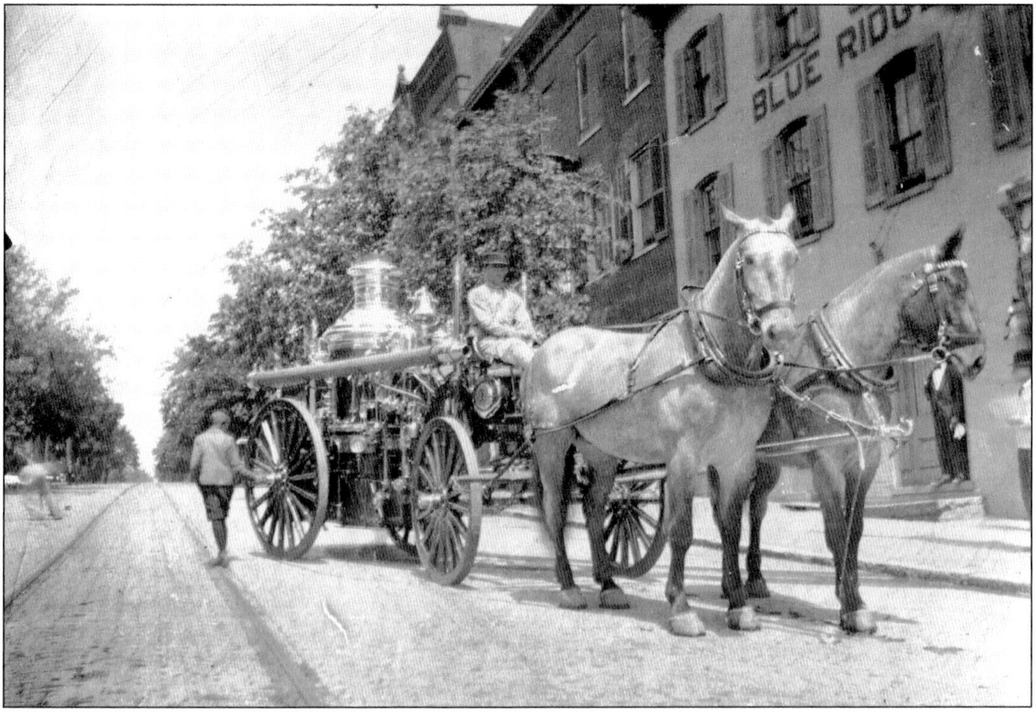

This view of the Junior's steamer was taken in front of the firehouse at 105 North Potomac Street. Note the trolley tracks running down the middle of the street. (Courtesy of the Independent Junior Fire Company.)

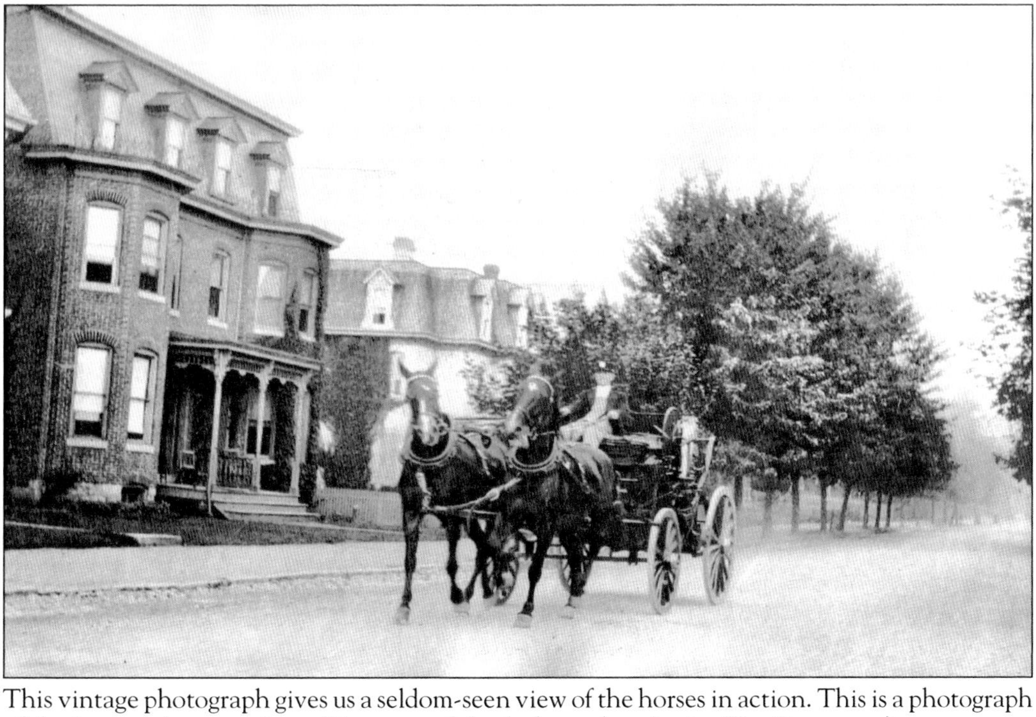

This vintage photograph gives us a seldom-seen view of the horses in action. This is a photograph of the Junior's hose carriage. (Courtesy of the Independent Junior Fire Company.)

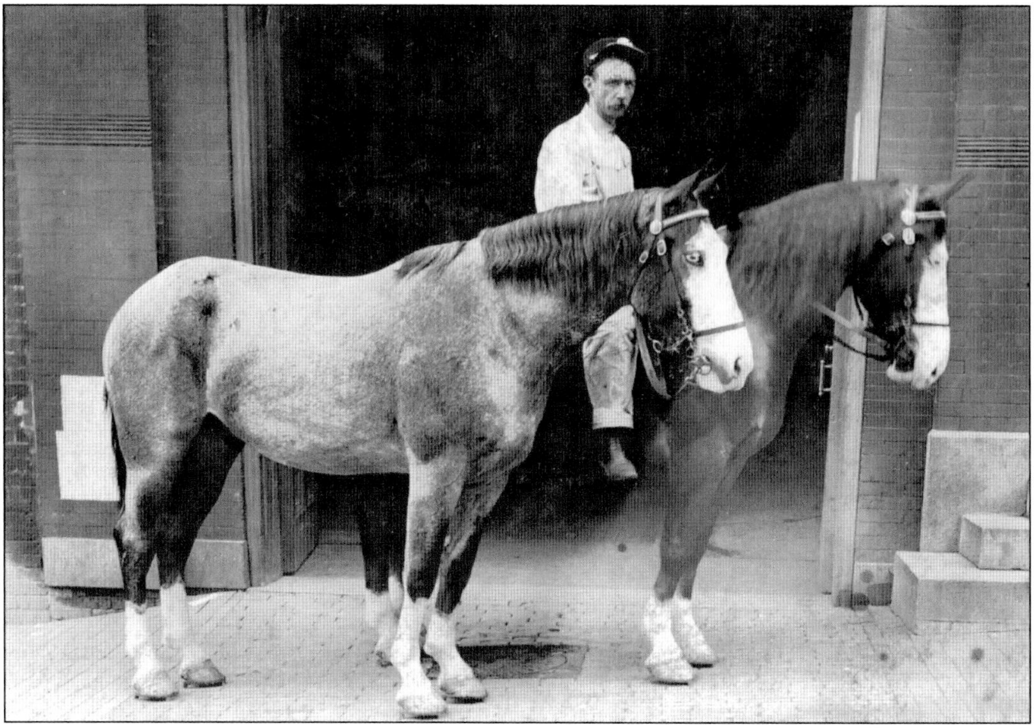

On March 8, 1908, the Juniors purchased these two fire horses. One is a bay and the other is a roan. The horses were purchased from Ott Smith, who lived one mile northwest of Leitersburg, Maryland. The horses weighed 2,660 pounds. One has two glass eyes and one has one glass eye. They were full brothers and were bred in Virginia. Standing fifteen-hands and three-inches tall, they cost $450. The Juniors sold the old horses for $225. (Courtesy of the Independent Junior Fire Company.)

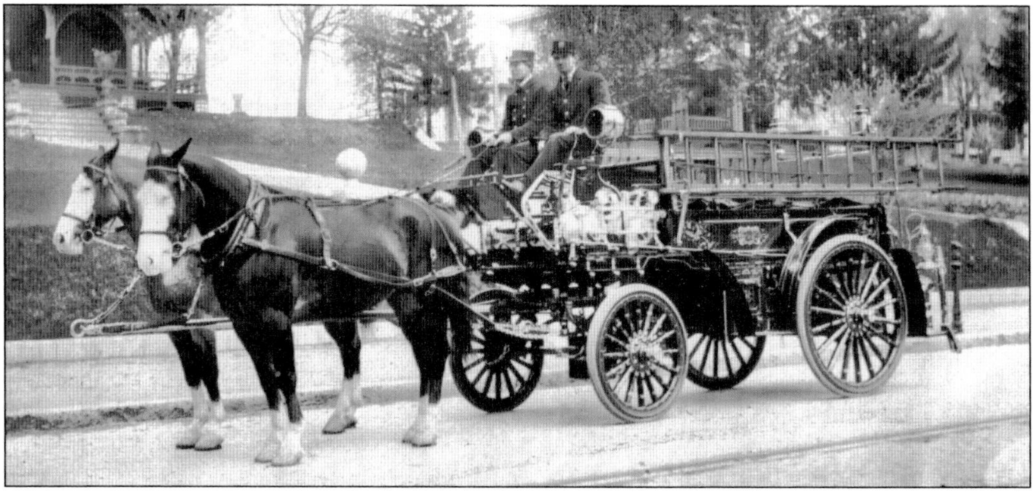

In the late 1800s and early 1900s, combination chemical/hose wagons were popular. When sulfuric acid and bicarbonate of soda (baking soda) are combined, an immediate chemical reaction takes place, creating carbon-dioxide gas. The pressure that is created forces the soda water or "chemicals" onto the fire. (Information courtesy of *Chemical Fire Engines* by W. Fred Conway.) (Courtesy of the Independent Junior Fire Company.)

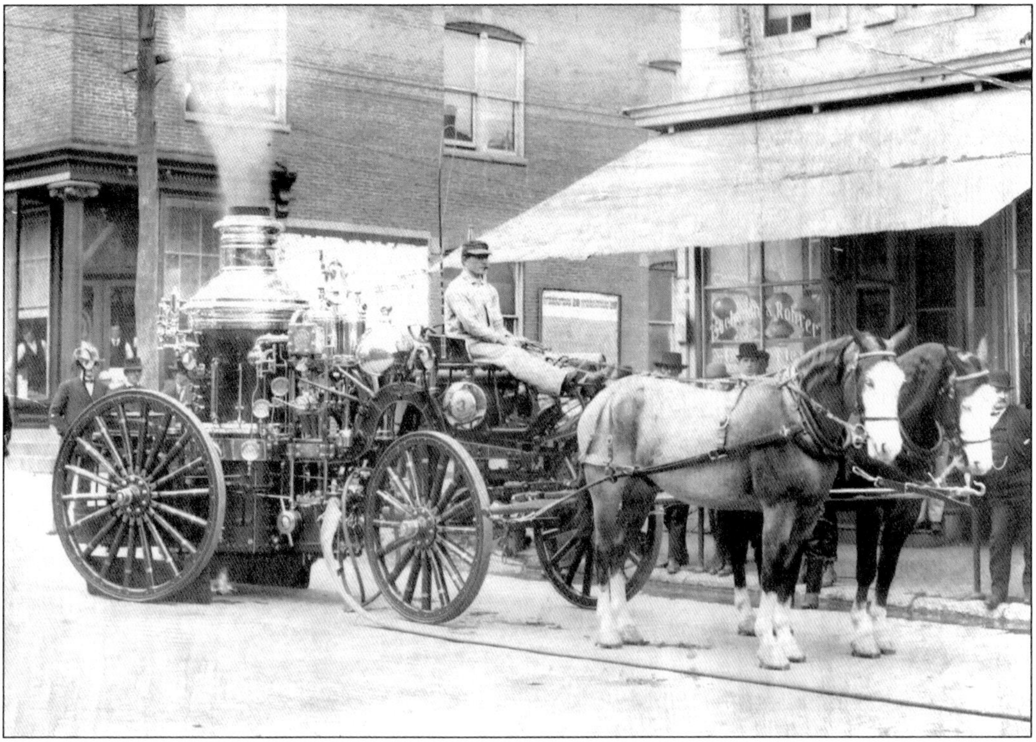

This rare photograph of the Junior's steamer making steam was taken at the corner of South Potomac and Antietam Streets in 1908. (Courtesy of the Independent Junior Fire Company.)

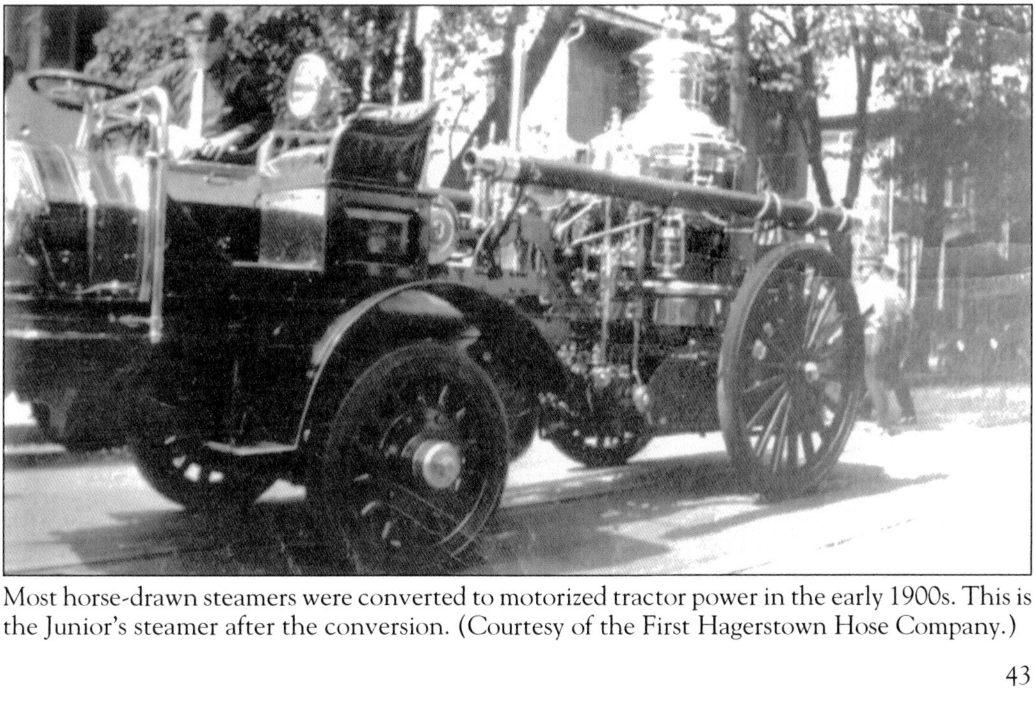

Most horse-drawn steamers were converted to motorized tractor power in the early 1900s. This is the Junior's steamer after the conversion. (Courtesy of the First Hagerstown Hose Company.)

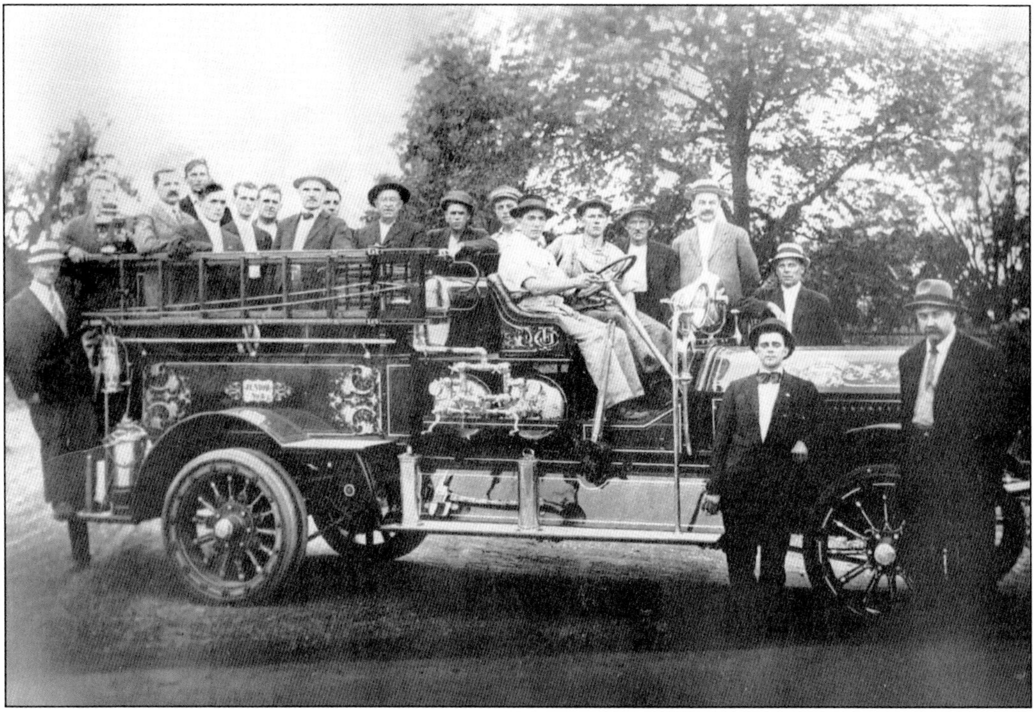

This 1915 chemical engine was delivered on July 10, 1915. The chemical tanks were replaced by a water pump in later years. Note the dress code of the era. The members are wearing dress coats, ties, and differently styled hats. (Courtesy of the Independent Junior Fire Company.)

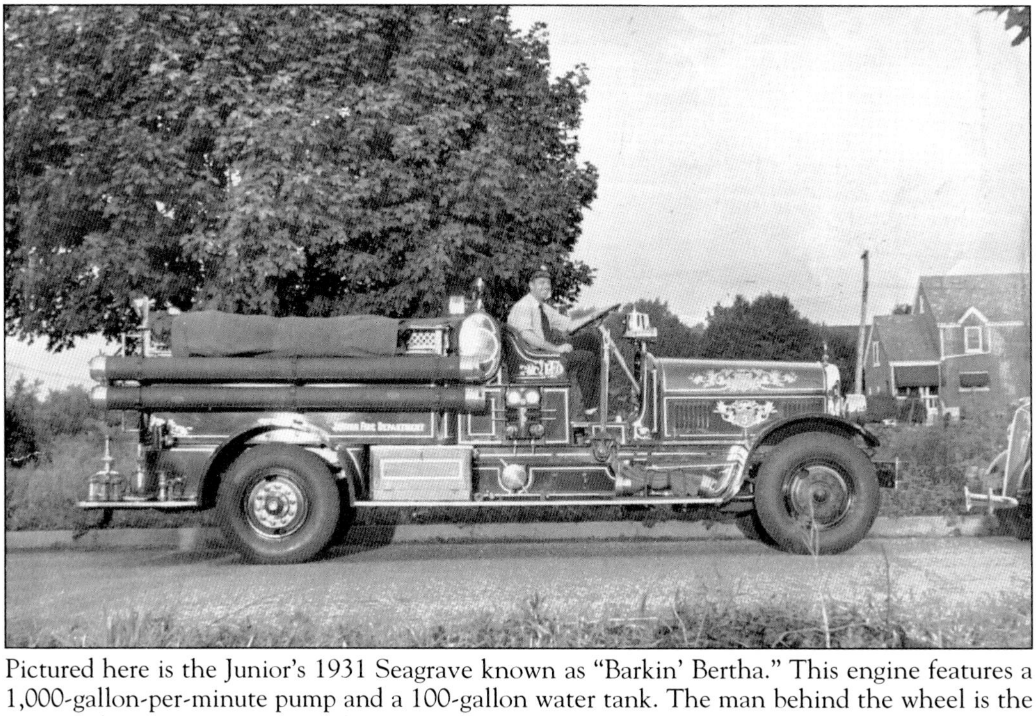

Pictured here is the Junior's 1931 Seagrave known as "Barkin' Bertha." This engine features a 1,000-gallon-per-minute pump and a 100-gallon water tank. The man behind the wheel is the Junior's driver, George Basore. (Courtesy of the George Basore collection.)

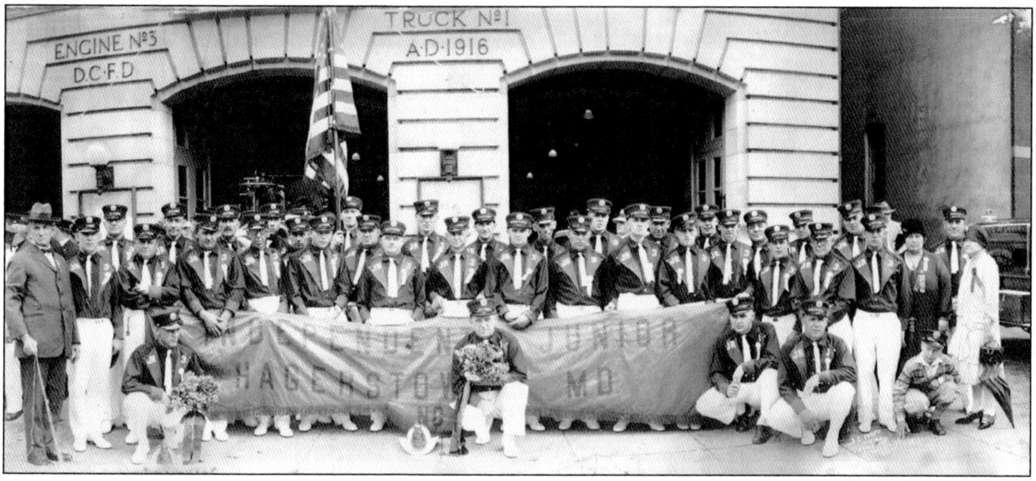

This photograph show the Juniors at a parade in Washington, D.C. The firehouse is similar to their own in the early 1900s. They were both the home of an Engine 3 and Truck 1. (Courtesy of the George Basore collection.)

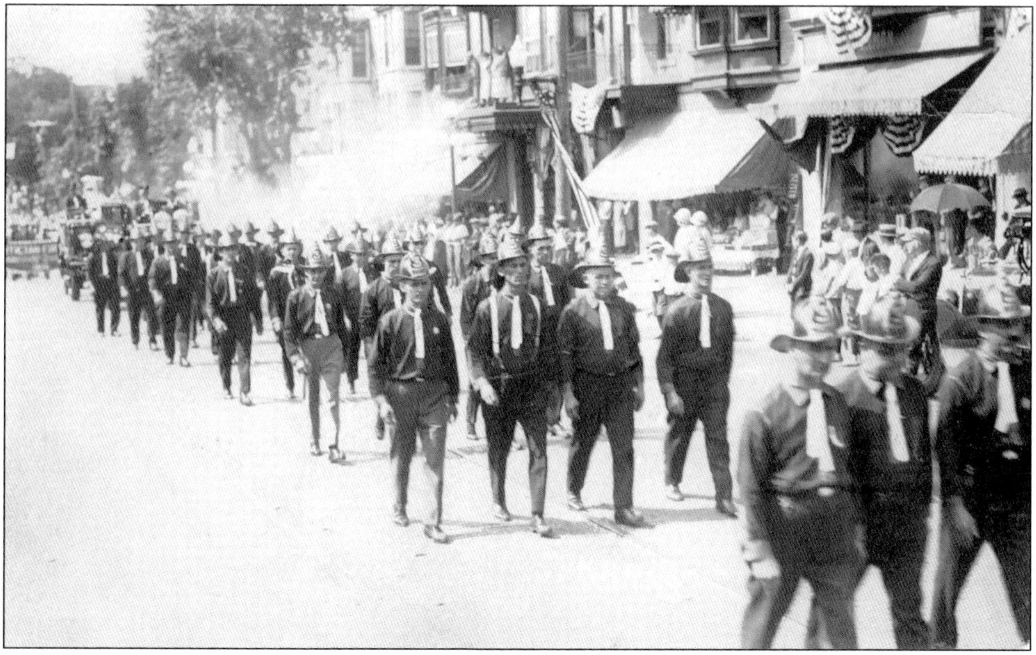

The attendance at parades gave the members a chance to show the citizens their pride in being firefighters. Putting your own life on the line to save others is a noble calling. (Courtesy of the Independent Junior Fire Company.)

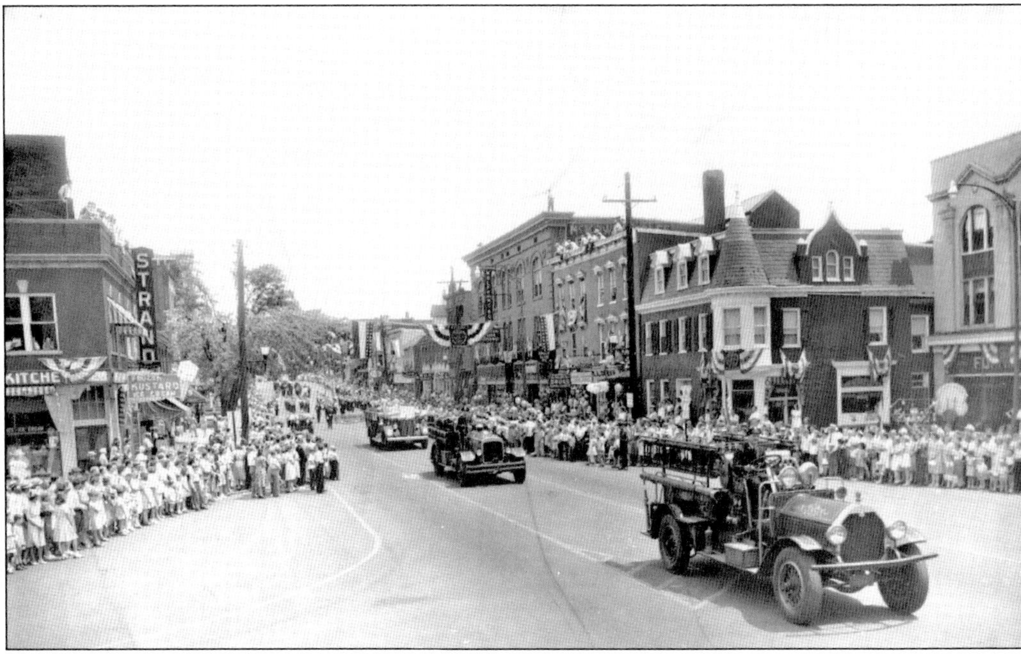

Between 1915 and 1965 the Juniors only used Seagrave fire engines. Shown here in a parade are the 1915, 1931, and 1947 Seagraves. (Courtesy of the George Basore collection.)

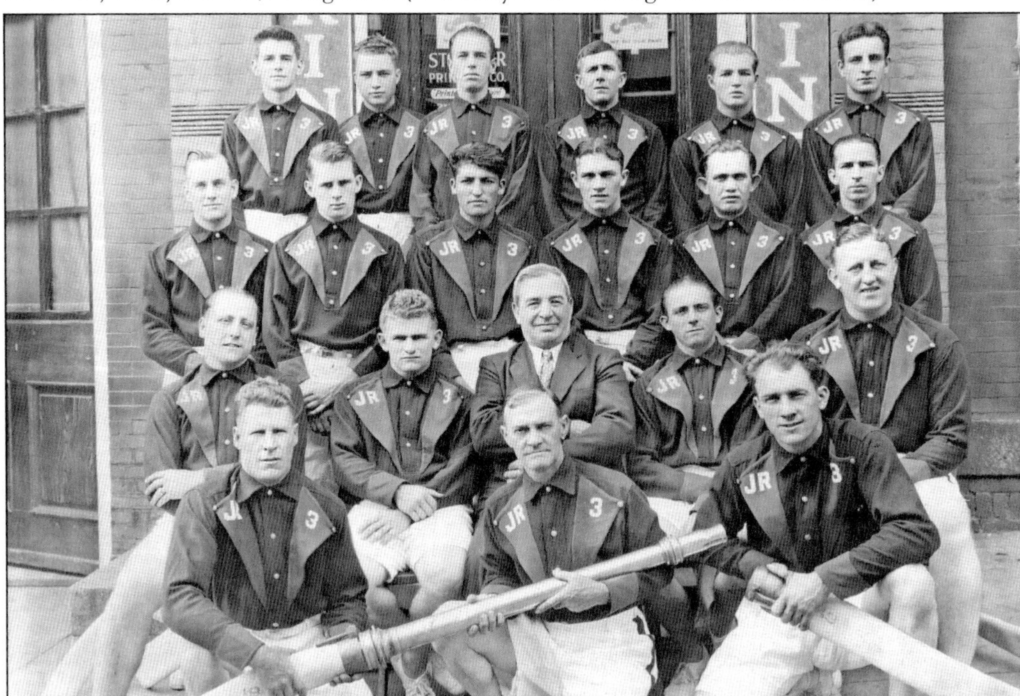

This is the Junior's hose-reel team in August 1934. The competition was in Waynesboro, Pennsylvania. The team had a time of 40 seconds, which was good enough for first place. (Courtesy of the George Basore collection.)

The Juniors occupied this site on North Potomac Street from 1852 until 1993. The new firehouse is located on North Eastern Boulevard. The building is currently used as a framing store. Legend has it that the drivers would wake up at night to the sound of the horses prancing just before the dispatch of a fire. (Courtesy of the First Hagerstown Hose Company.)

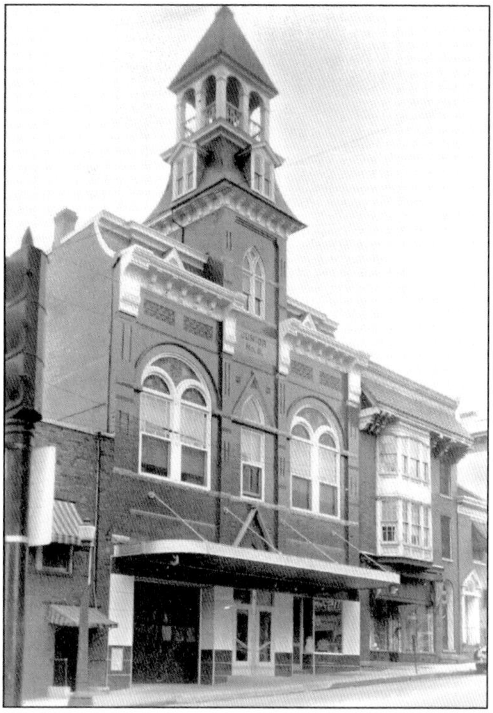

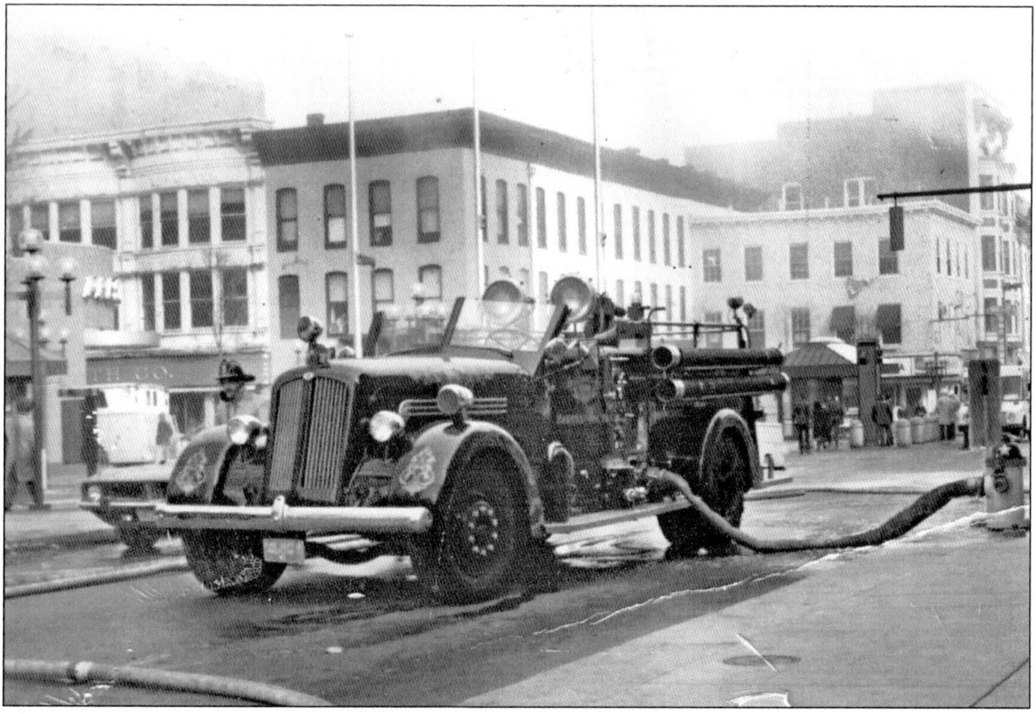

The Junior's 1947 Seagraves engine is seen pumping from the public square to the Glick Shoe Store fire on West Washington Street on February 1, 1977. The engine packed a V-12 motor and could pump 1,000 gallons per minute. The fire apparatus operator is James Stevens. FAO Stevens began his career on August 8, 1975. (Courtesy of the James Stevens collection.)

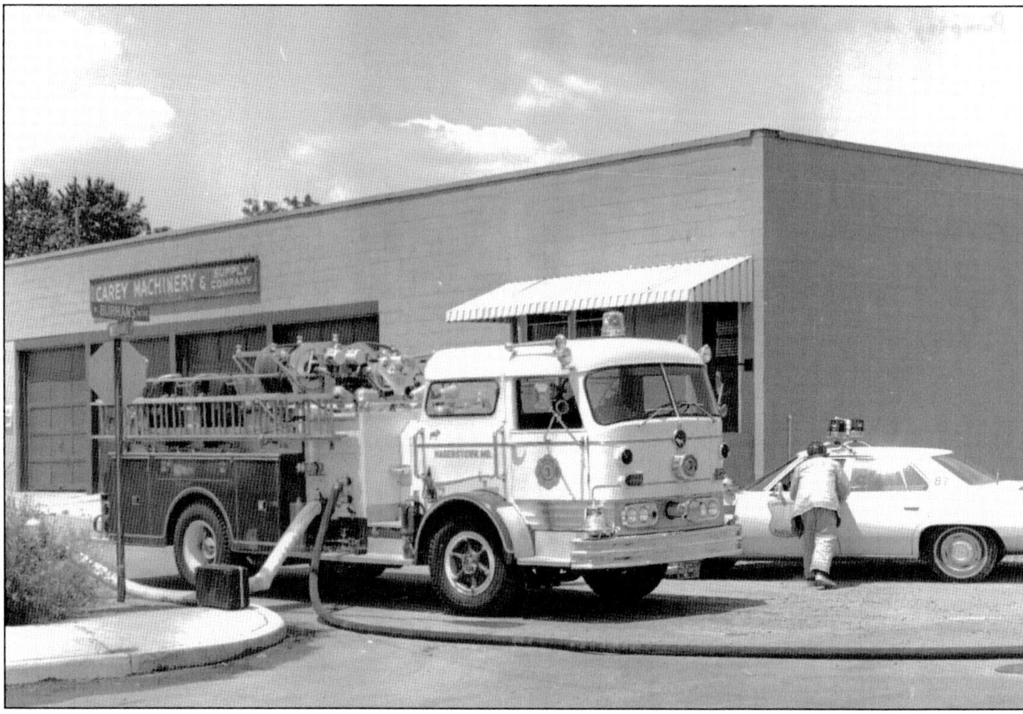

The Juniors replaced the 1947 Seagrave with this 1965 Mack engine. The Mack can pump 1,000 gallons per minute and has a 300-gallon water tank. On this day (June 5, 1976), the Mack is pumping from Mechanic Street to the Bester-Long fire on North Burhans Boulevard. This engine was dedicated to driver Benjamin Conrad for providing the city with over 33 years of dedicated service. (Courtesy of the First Hagerstown Hose Company.)

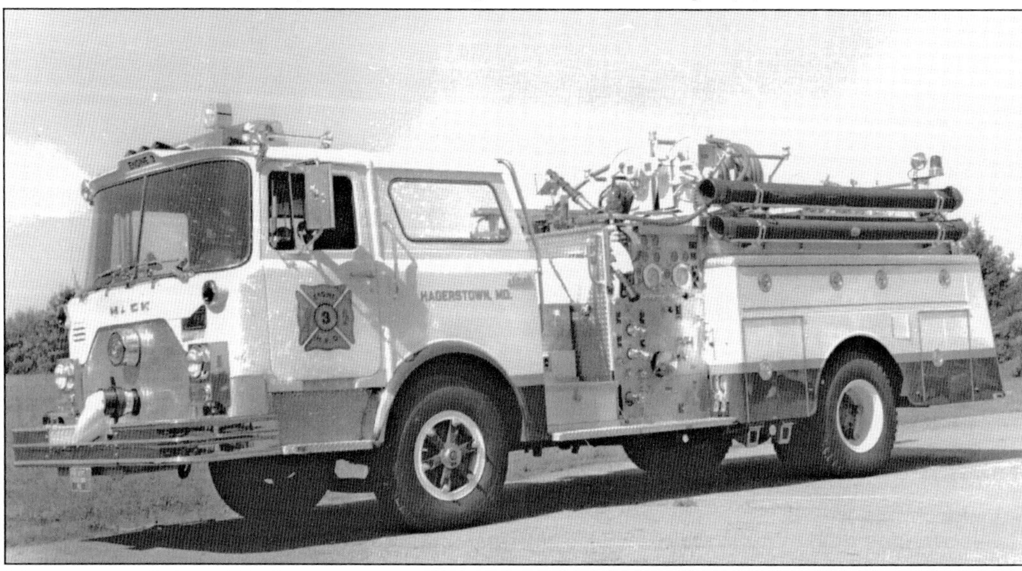

This is one of three Mack engines purchased at the same time in 1982. It has a 1,000-gallon-per-minute pump and a 500-gallon water tank. The engine was replaced in February 2004 with a new KME engine. The engine remains in service as a reserve engine. (Courtesy of the Rick Kipe collection.)

Four

Western Enterprise Fire Company, Engine-Truck 4

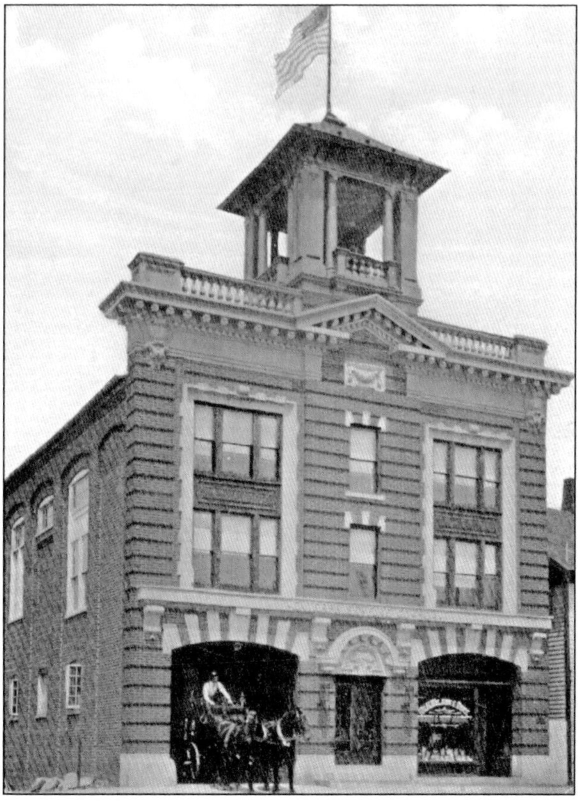

By 1869 the population in the western part of town began to increase to the extent that the three fire companies, all located in the lower end of town, were not adequate for the protection of the West End. A number of citizens from the West End incorporated the Western Enterprise Fire Company on September 19, 1872. This postcard view shows the horses coming out of the Enterprise's second permanent home located in Washington Square about 1908. (Courtesy of the David White collection.)

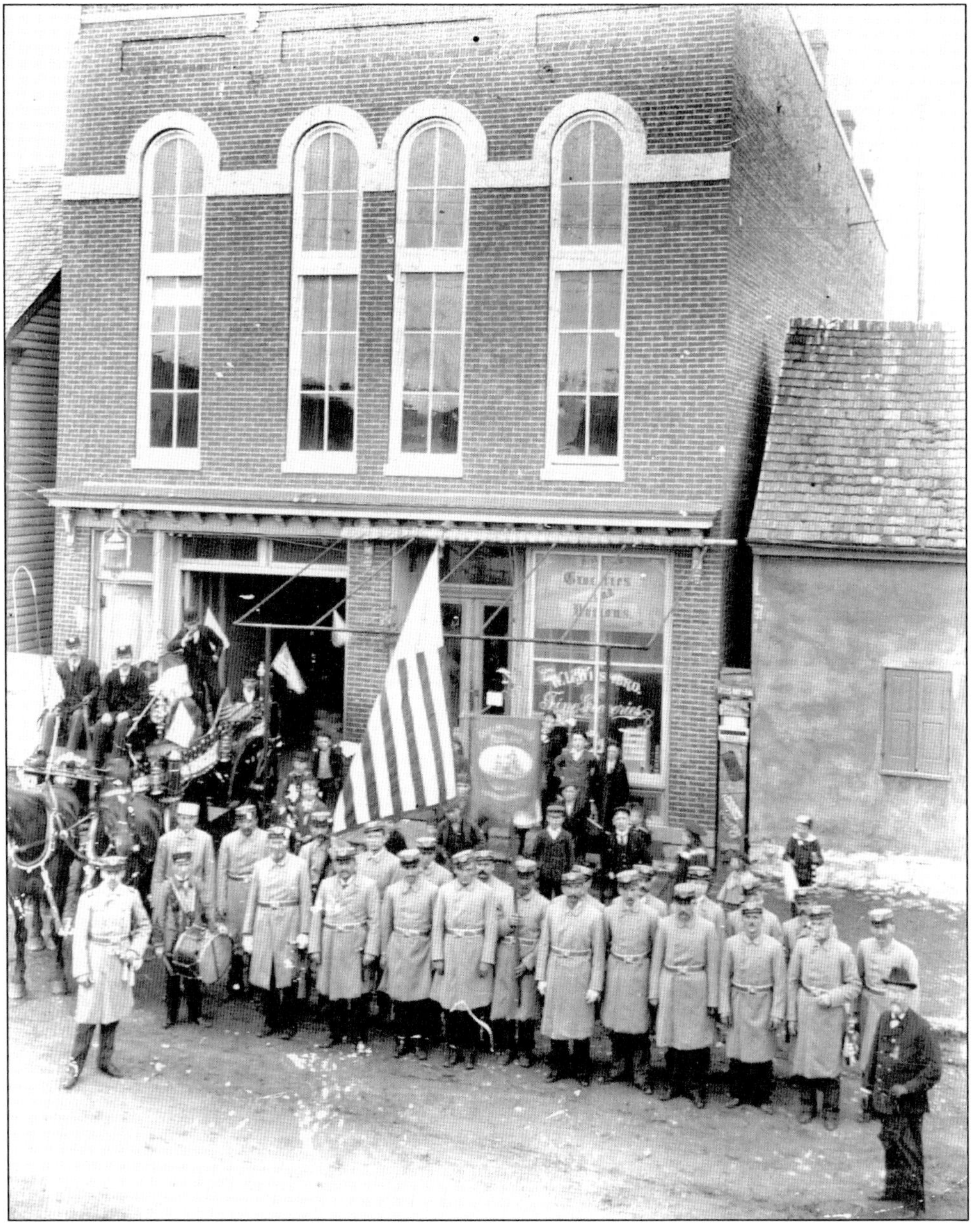

The Enterprise's first piece of apparatus was a hand pumper named "Big Six," which was donated by the Independent Junior Fire Company. Big Six was housed in a stable. Soon a movement was on to construct a firehouse. A frame building was erected on the site of the old Steam Engine Works property. After a few years, the site was needed by the owners so the Enterprise moved to a lot on the American House property. This only lasted a short while and another move was necessary. The Enterprise finally secured a lot on West Franklin Street (the present-day 300-block near the railroad overpass on the south side of the street). This is the only known photograph of the Franklin Street Firehouse. It was a two-story brick building that contained a hall on the second floor and a meeting room, engine room, and a store room to be rented on the first floor. It cost less than $6,000 to build. By 1894 the Enterprise owned a steamer, hose carriage, and a hose cart. (Courtesy of the Western Enterprise Fire Company.)

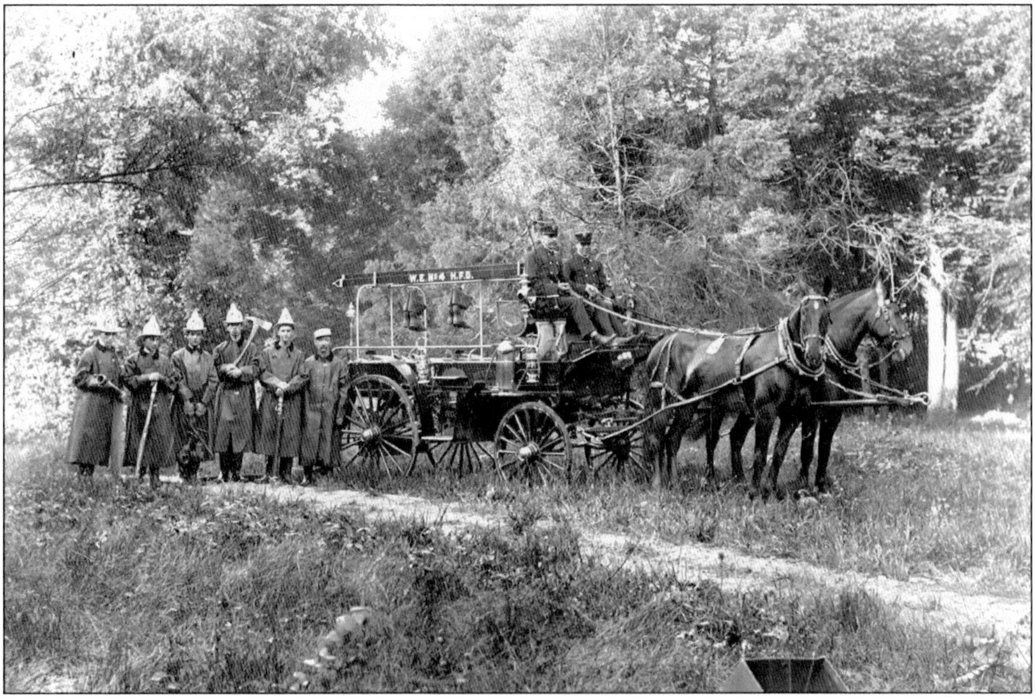

This horse-drawn hose wagon is the same piece of apparatus shown on the previous page. Most likely several of the same firefighters are in both photographs. Note the equipment the men are displaying. (Courtesy of the Western Enterprise Fire Company.)

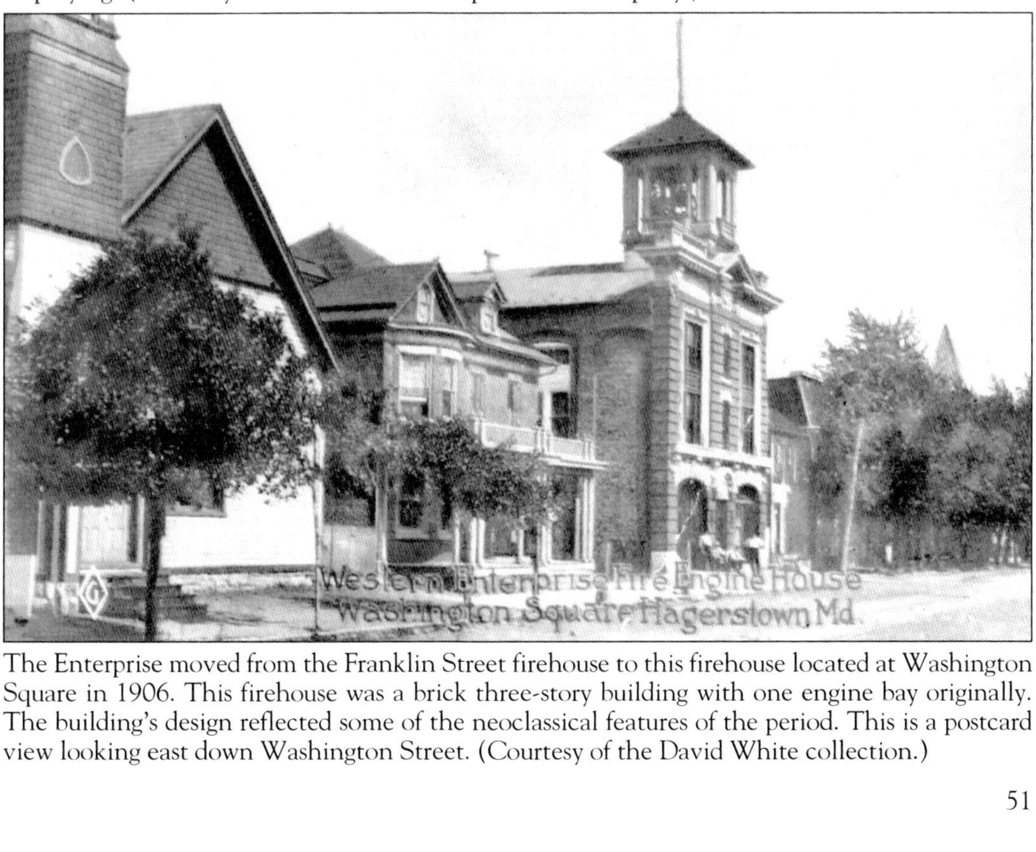

The Enterprise moved from the Franklin Street firehouse to this firehouse located at Washington Square in 1906. This firehouse was a brick three-story building with one engine bay originally. The building's design reflected some of the neoclassical features of the period. This is a postcard view looking east down Washington Street. (Courtesy of the David White collection.)

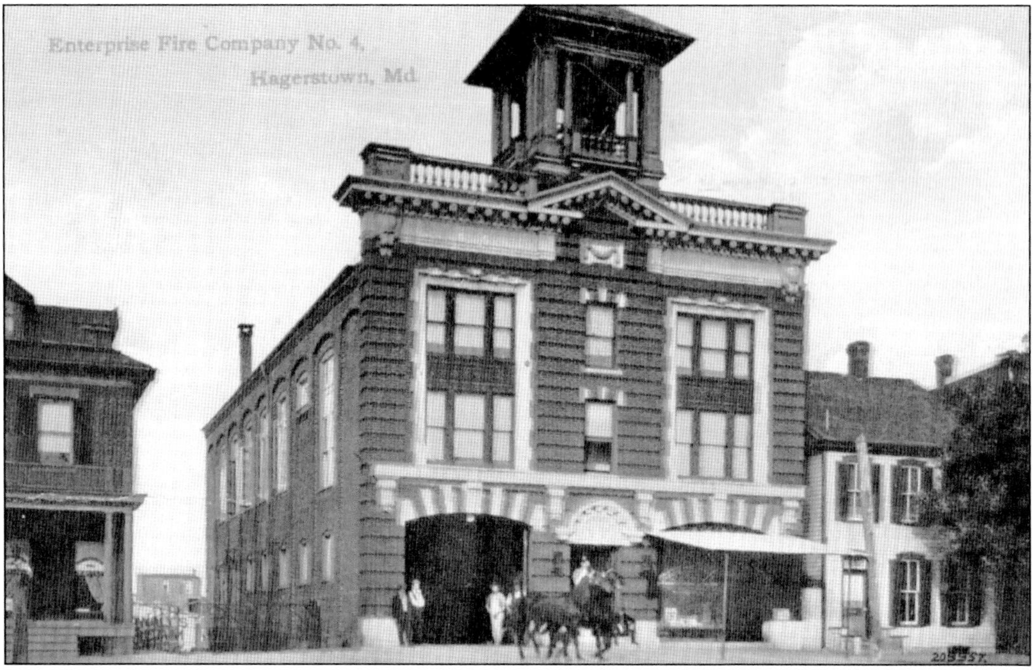

The fire horses were taken for a walk each day at 5:30 p.m. for exercise. In this photograph it appears the horses are taking their daily stroll. This postcard is dated 1911. (Courtesy of the David White collection.)

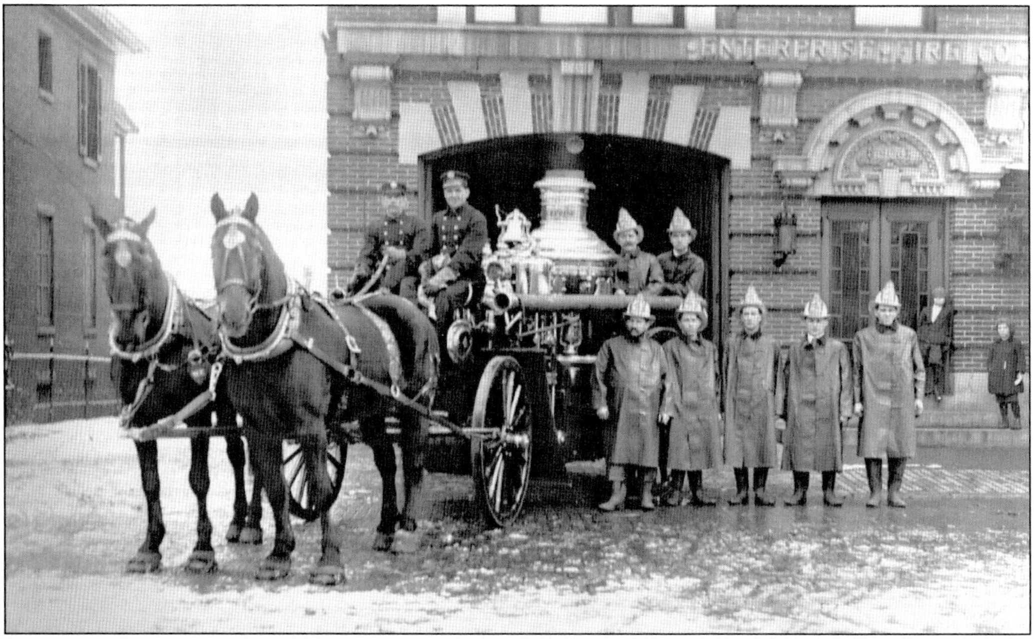

It appears the members of the Enterprise are ready to fight fire. One of the drivers in this picture is Harry Borne; the other is unidentified. From left to right are (front row) ? Myers, Ralph Grove, Frank Golden, Clarence Brewer, and Edgar Boone; (back row) Dodger Sheetz and John Fisher. The horses are Snap and Freddy. Note the horses are wearing chains on their hooves and a dog is resting on the lap of the driver. (Courtesy of the George Basore collection.)

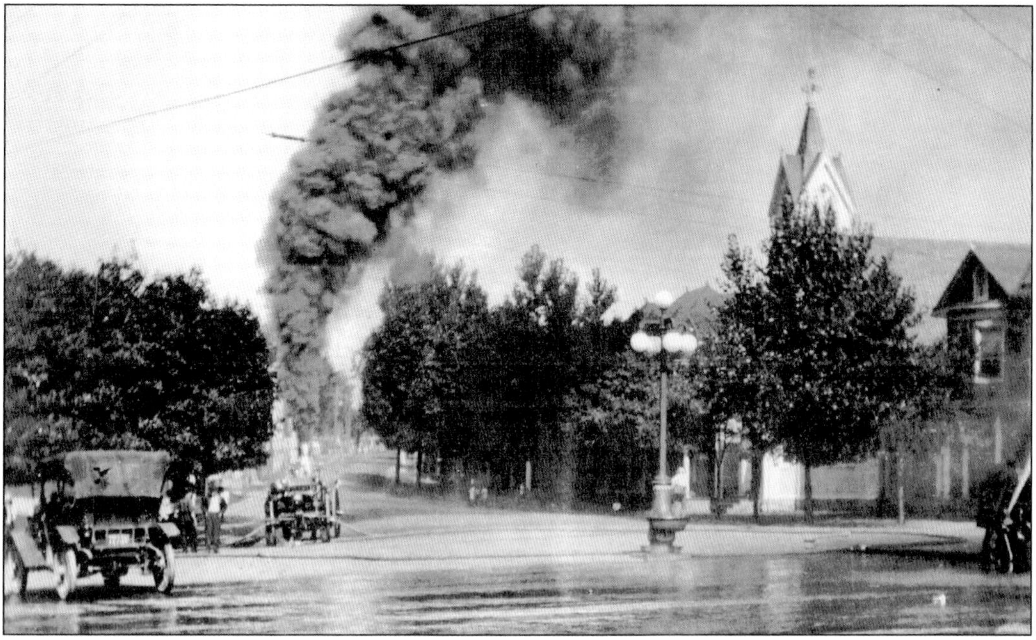

The Enterprise steamer is pumping from a cistern at a drill or test. This photograph was taken in Washington Square in the late 1910s or early 1920s. The firehouse is just out of view on the right. The last fire this steamer pumped was at the Westside Lumber Company in about 1929 or 1930. (Courtesy of the William Dieterich collection.)

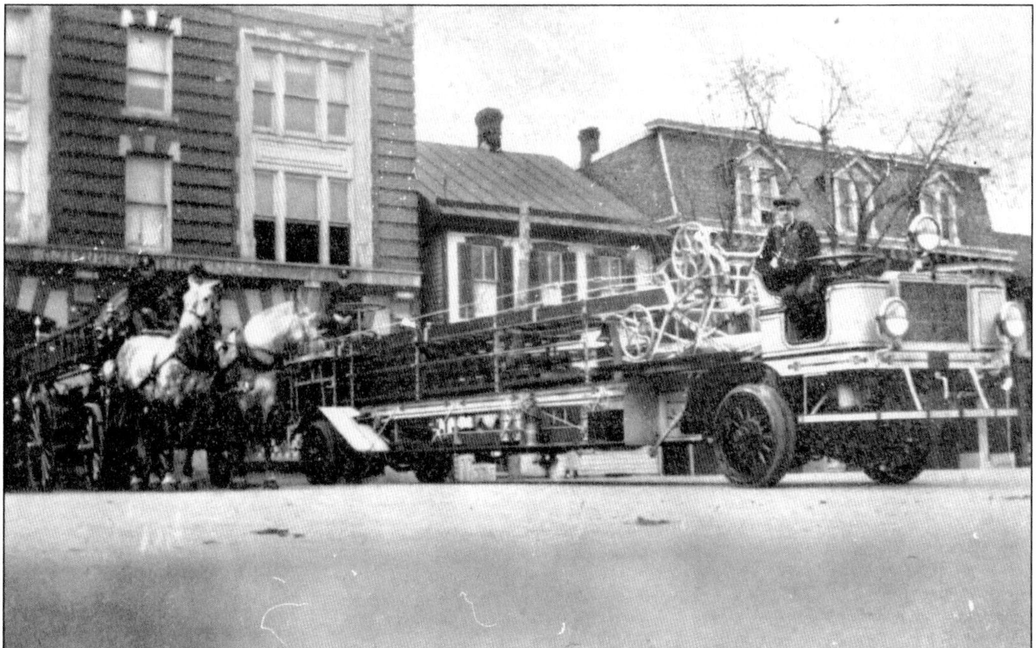

With the advent of motorized apparatus, the days of horse-drawn apparatus came to an end. In this photograph is the Pioneer Hook & Ladder's 1914 American LaFrance ladder truck parked next to a Western Enterprise hose wagon. Note the horse on the right checking out the motorized truck. (Courtesy of the William Dieterich collection.)

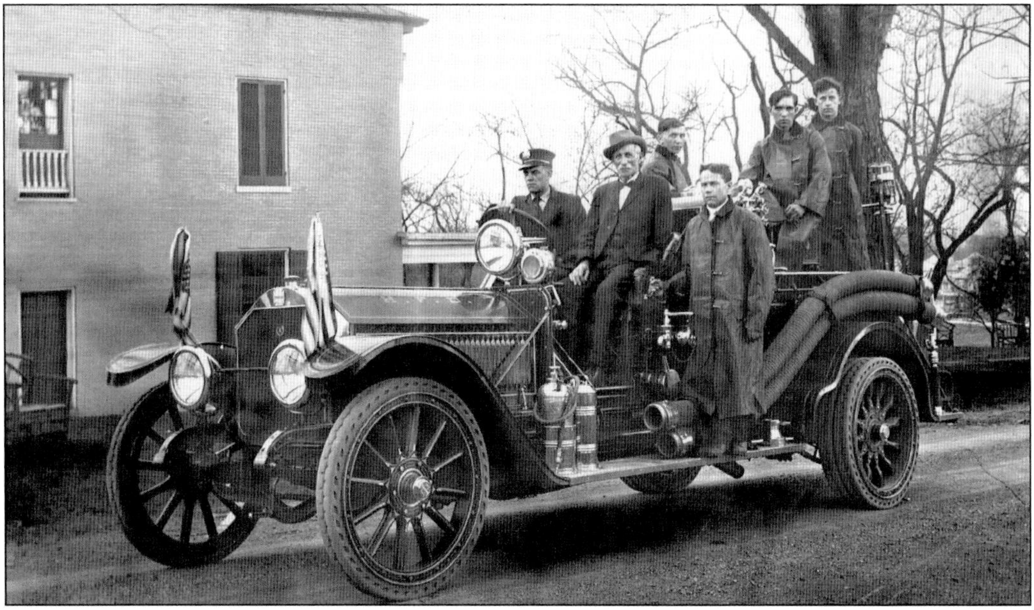

This photograph of an early-model American LaFrance pumper with firefighters aboard was taken in front of the Mansion House at City Park in the early 1920s. From left to right are Driver Freaner Logan, Jake Baumgart, Fred Wellinger, George Alexander, unidentified, and Charles Goetz. The pumper was designed to replace the horse-drawn apparatus. (Courtesy of the Western Enterprise Fire Company.)

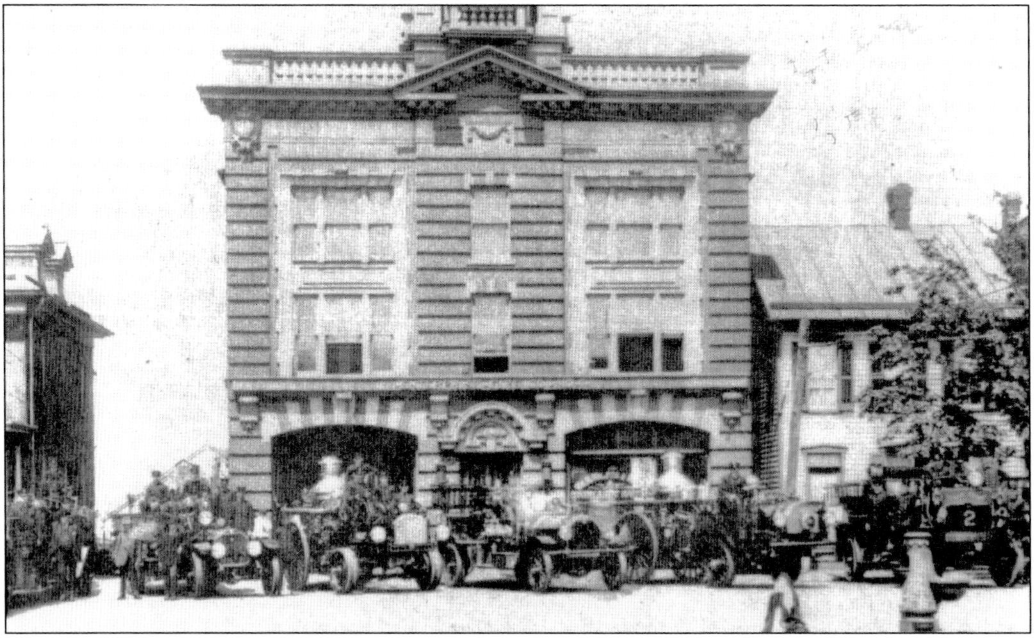

By 1918 each city firehouse had at least one piece of motorized fire apparatus. Represented here from left to right are Pioneer Hook & Ladder Company's 1914 American LaFrance ladder truck, Western Enterprise's American LaFrance pumper and steam engine, Independent Junior's 1915 Seagrave pumper and steam engine, and the Antietam Fire Company's 1913 Knox pumper. (Courtesy of the David White collection.)

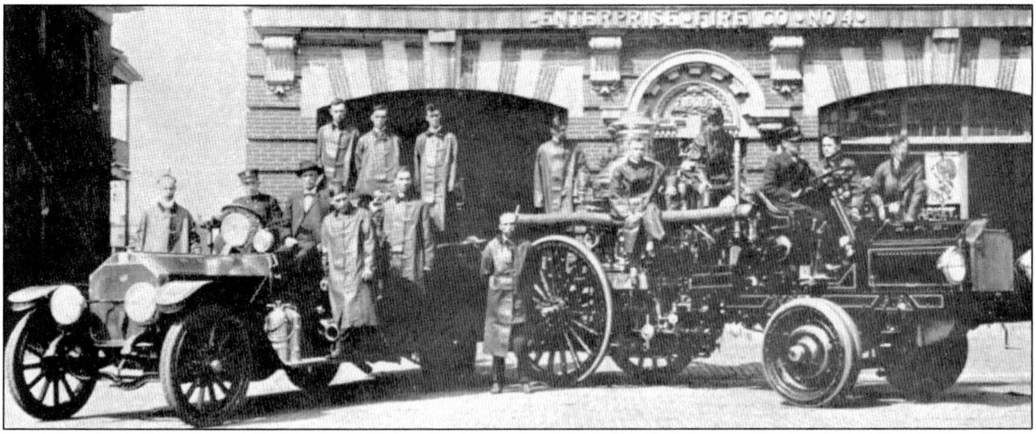

This is a close-up view of the Enterprise's first two pieces of motorized fire apparatus. The steamer was originally horse-drawn but was converted to motorized. The active members are well represented in this photograph. (Courtesy of "1936 Maryland State Firemen's Convention.")

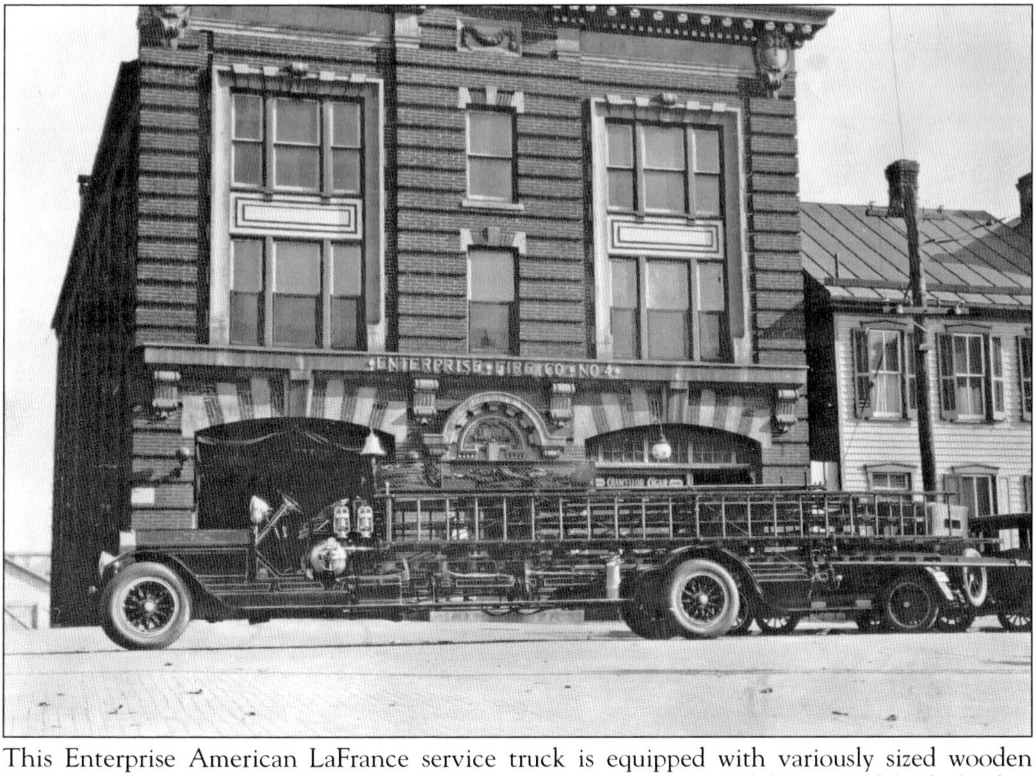

This Enterprise American LaFrance service truck is equipped with variously sized wooden ground ladders and a chemical tank. This is considered a service truck because of the lack of an aerial ladder. (Courtesy of the Jo Ann Kendall collection.)

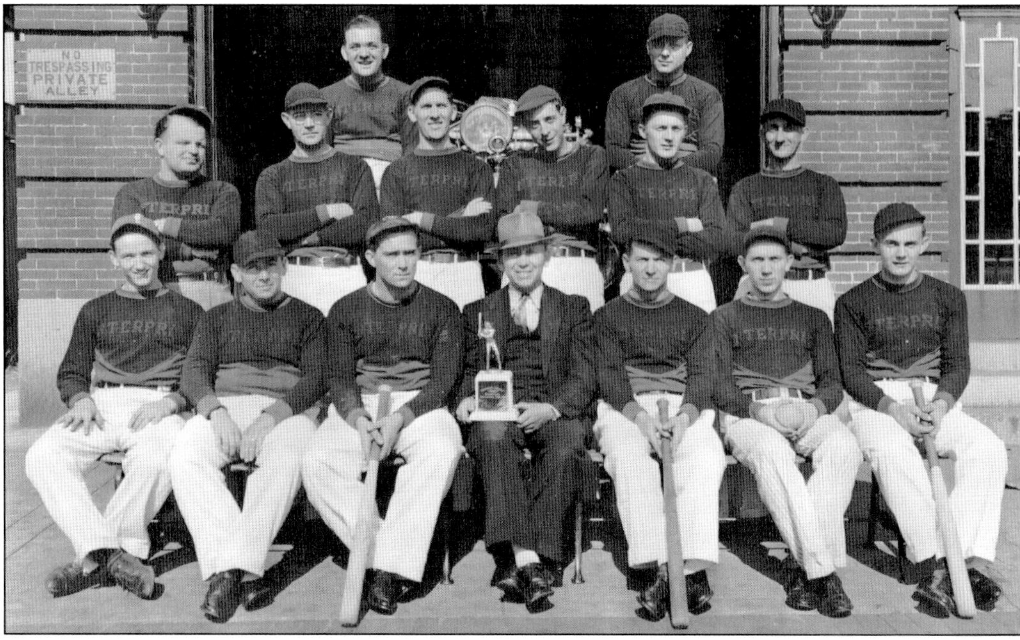

Many fire companies during this era had softball teams. This Enterprise team won the 1938 fast-pitch softball title. From left to right are (front row) Johnny Burger, unidentified, Carl Smith, two unidentified people, Al Burger (pitcher), and Bob Swartz; (middle row) Jim Dixon (catcher), Charles "Ticker" Athey, Wilber "Booby" Cousins (second base), Clinton "Fats" Lapole, Pete Limburg (third base), and Charlie Socks; (back row) Irvin Cousins and ? Swartz. (Courtesy of the Western Enterprise Fire Company.)

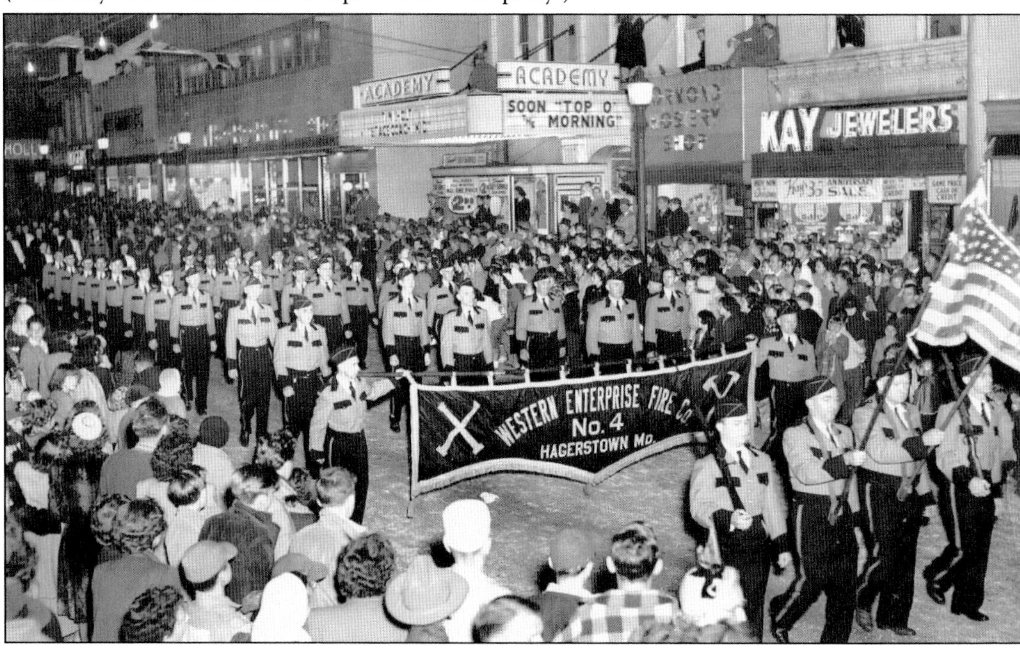

In this photograph, the Western Enterprise drill team marches in the Alsatia Mummers parade in the late 1940s. The uniforms were orange with black trim and members purchased their own. (Courtesy of the Terry Mose collection.)

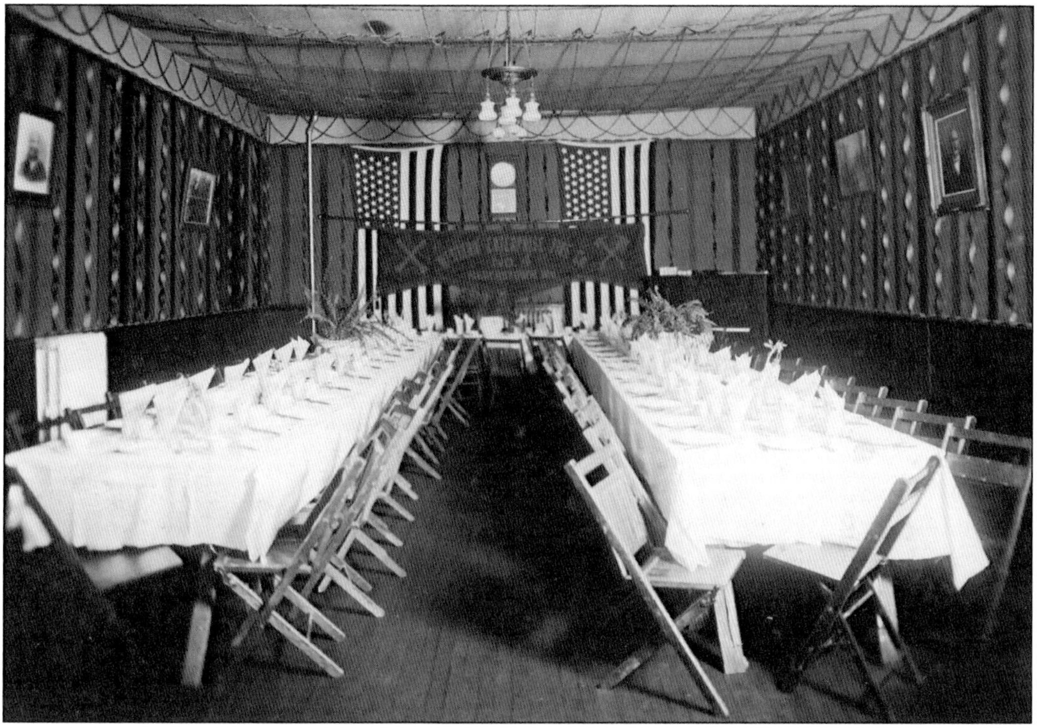

This is an inside view of the Western Enterprise hall. The room is set up for a Machinist Union function. Many fire company members worked for the railroad. (Courtesy of the Western Enterprise Fire Company.)

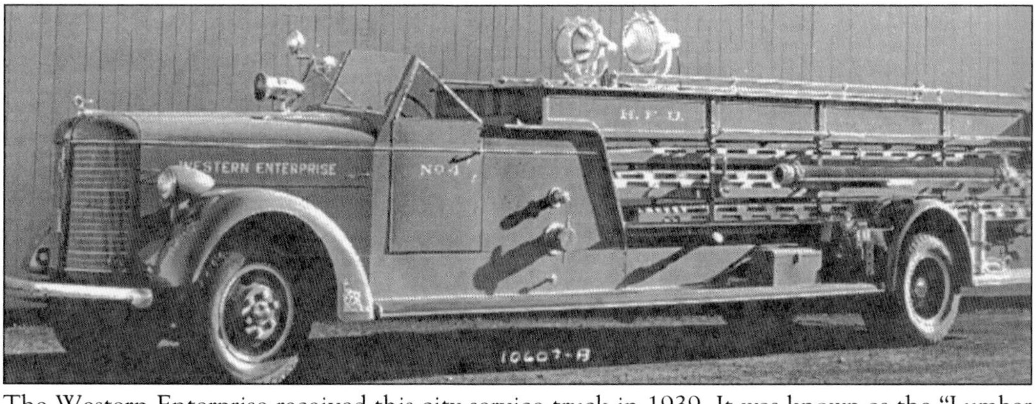

The Western Enterprise received this city service truck in 1939. It was known as the "Lumber Pile" because of the numerous wooden ladders that protruded from the rear. It is an American LaFrance 500-series, type-945 open-cab quad. This is the factory photograph. (Courtesy of *The Encyclopedia of American LaFrance*.)

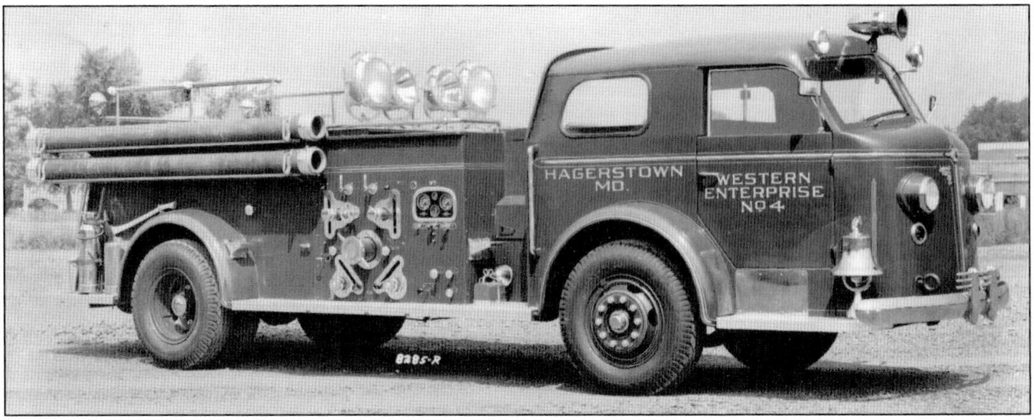

This is the factory photograph of the Western Enterprise 1948 American LaFrance pumper. It has a 1,500-gallon-per-minute pump and a 300-gallon water tank. (Courtesy of the JoAnn Kendall collection.)

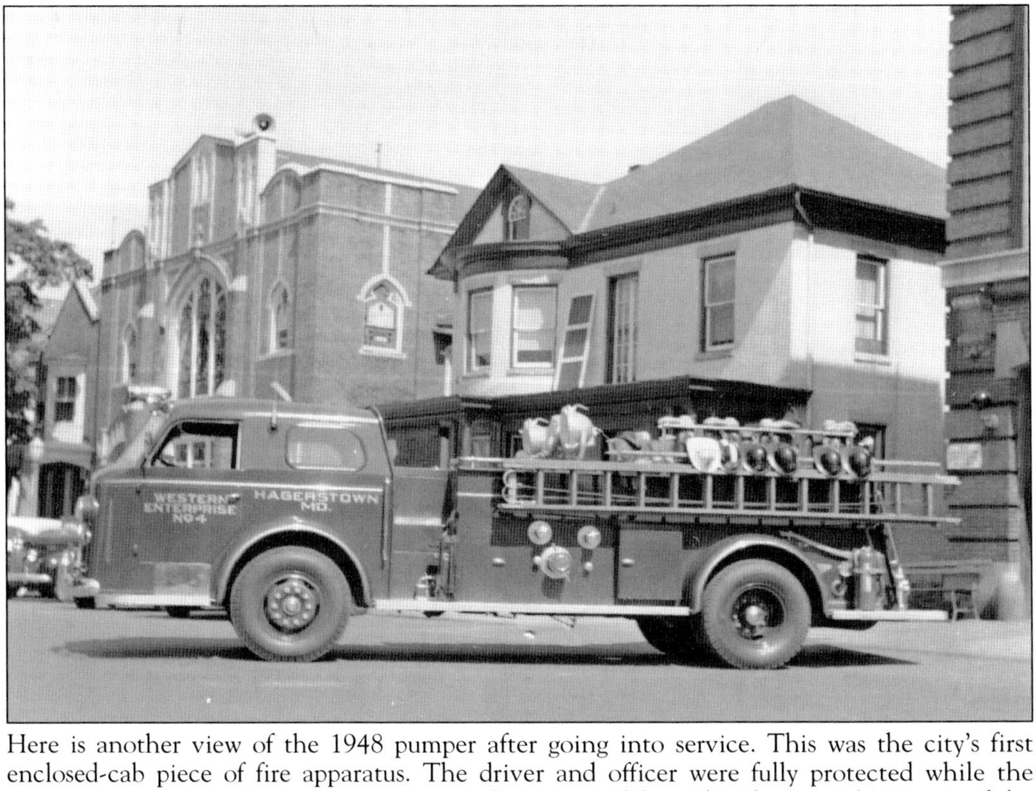

Here is another view of the 1948 pumper after going into service. This was the city's first enclosed-cab piece of fire apparatus. The driver and officer were fully protected while the firefighters in the jump-seat area were partially protected from the elements. (Courtesy of the C. William Karn collection.)

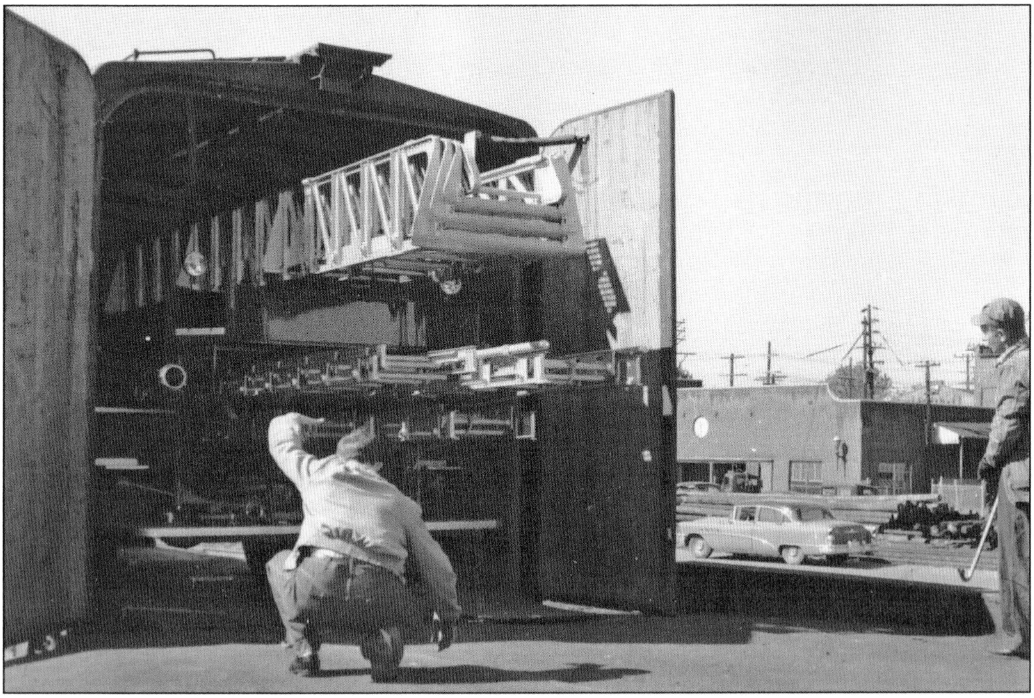

This is a picture of delivery day for the Western Enterprise 1956 Pirsch ladder truck, affectionately known as "Bruno." Legend has it that Bruno did not run off of gasoline. All that was required was a bowl of Alpo and a can of spring water. Bruno has a 75-foot aerial, 750-gallon-per-minute pump and a 200-gallon water tank. (Courtesy of the C. William Karn collection.)

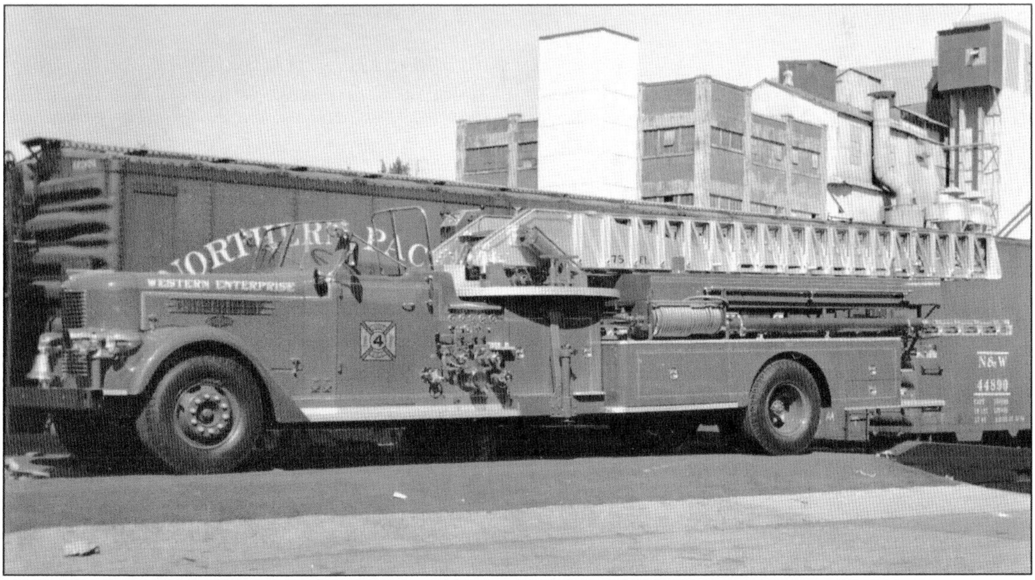

Fifty years ago it was not uncommon to have fire apparatus delivered by rail. "Bruno" was delivered at an unloading area off West Lee Street. Bruno served the city faithfully as a first-due truck until 1978. Then Bruno was leased and later sold to the Longmeadow Volunteer Fire Company for $1. Longmeadow operated the truck for another 14 years. (Courtesy of the C. William Karn collection.)

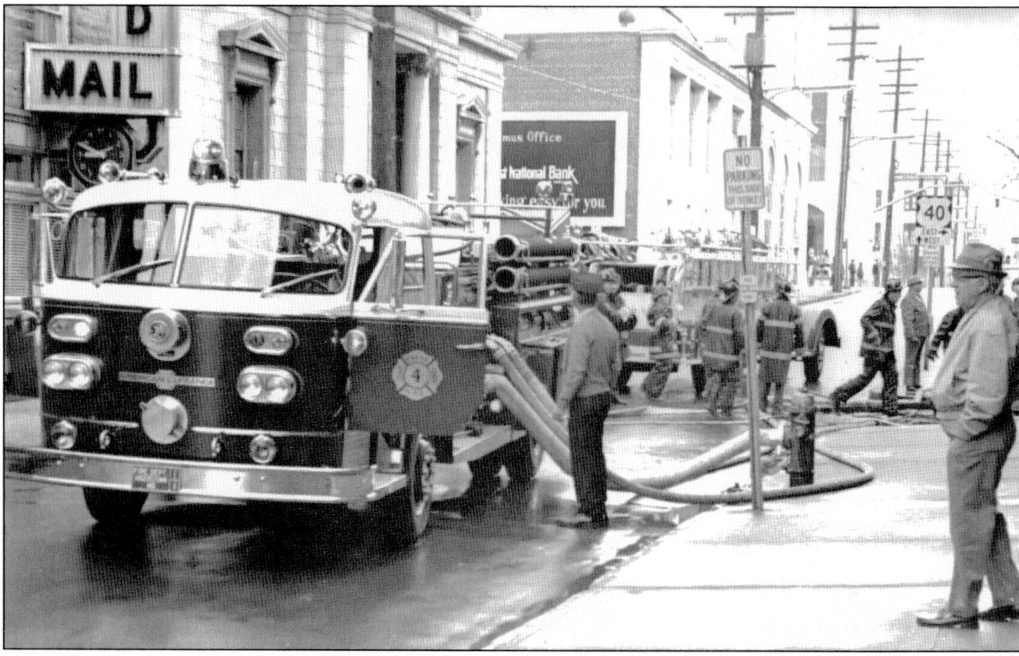

Water is being pumped to the Newberry's department store fire in 1974 with this 1969 American LaFrance pumper. It had a 1,000-gallon-per-minute pump and a 300-gallon water tank. The photograph was taken in the unit block of Summit Avenue. (Courtesy of the First Hagerstown Hose Company.)

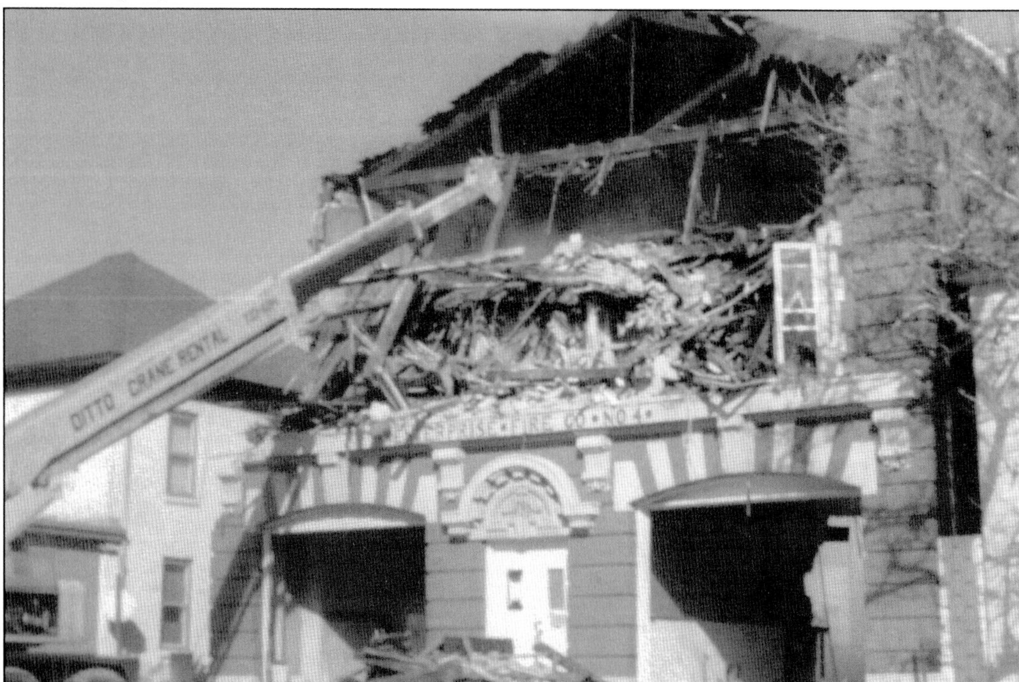

In 1984, the Enterprise demolished the 1906 firehouse and constructed a modern firehouse, which still exists on the site today. (Courtesy of the Jeff Snyder collection.)

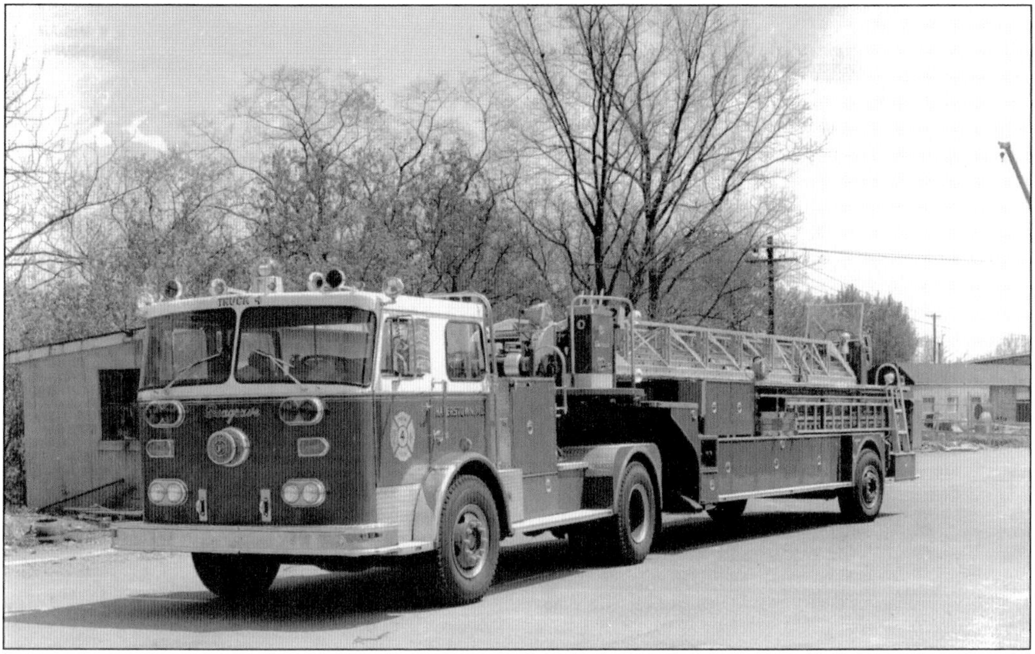

In 1978 the city purchased this Seagrave aerial ladder truck. The truck is equipped with a 100-foot aerial and a full compliment of ground ladders. The photograph was taken at the Hagerstown Fire Department Training Center off Bowman Avenue. The truck is currently in the Baltimore Fire Department fleet. (Courtesy of the Rick Kipe collection.)

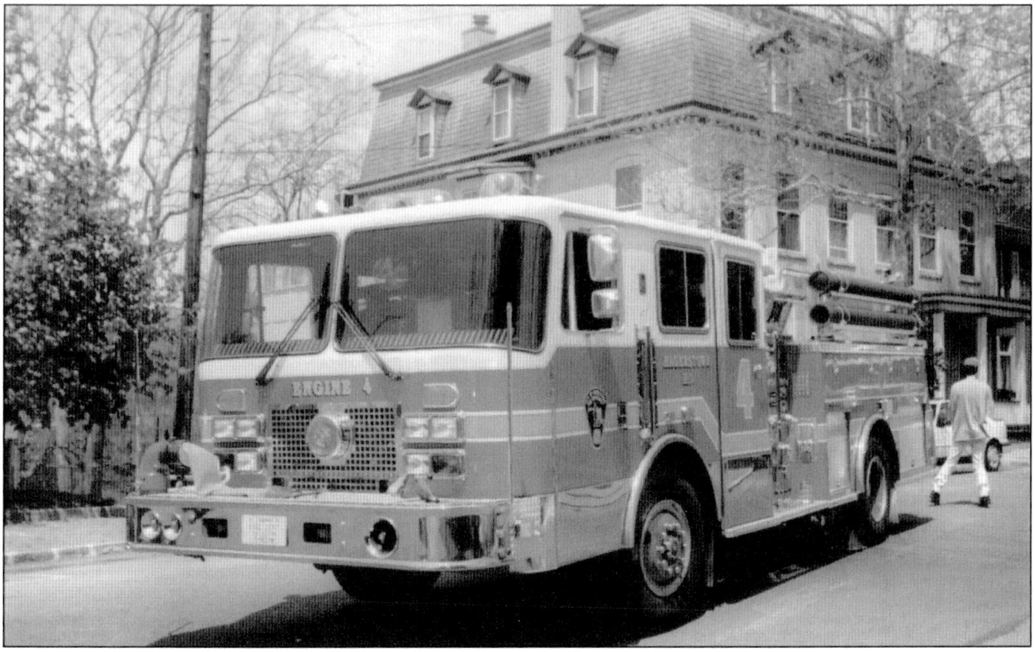

The first pumper that was not an American LaFrance in Western Enterprise history is this 1991 KME pumper. It has a 1,500-gallon-per-minute pump and a 500-gallon water tank. In the background is the building where firefighter Walter Sharar lost his life from a fall that occurred while battling a house fire in 1932. (Courtesy of the author.)

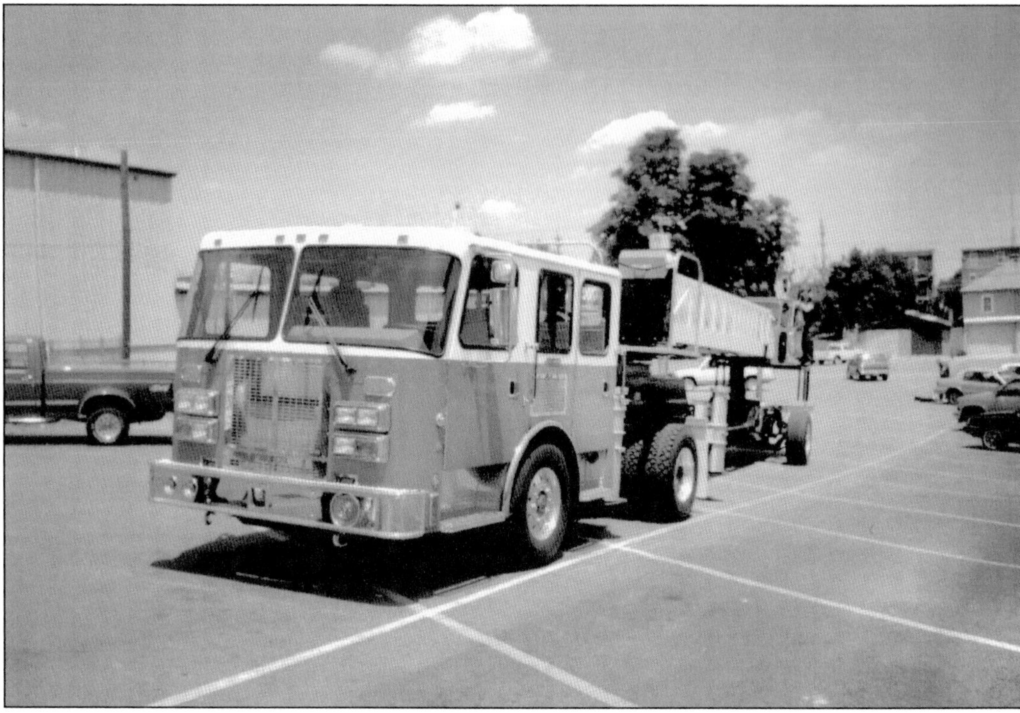

What is wrong with this picture? This is the new 1997 Quality/Simon-Duplex aerial ladder truck on its way through Hagerstown for final assembly in Alabama. (Courtesy of the author.)

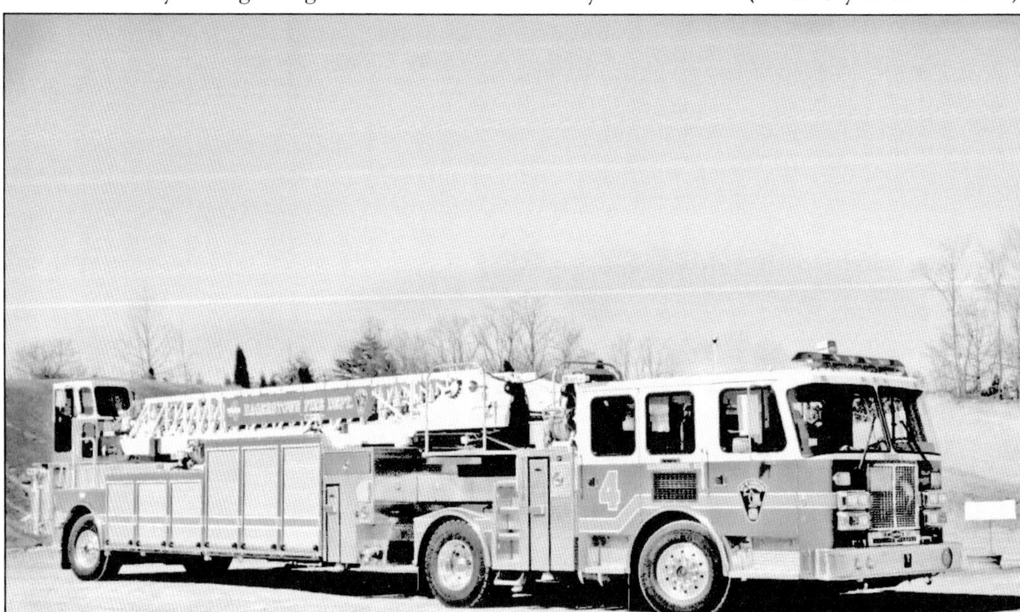

At a price tag of $549,298, this is the completed Western Enterprise Truck 4. Some of the best innovations in the past 25 years include improvements made to aerial apparatus. The aerial can be moved an inch at a time, if necessary, which helps the operator feather the aerial into tight places. This truck is a twin to the current Pioneer Hook & Ladder's Truck 1. (Courtesy of the Warren Jenkins collection.)

Five

South Hagerstown Fire Company, Engine 5

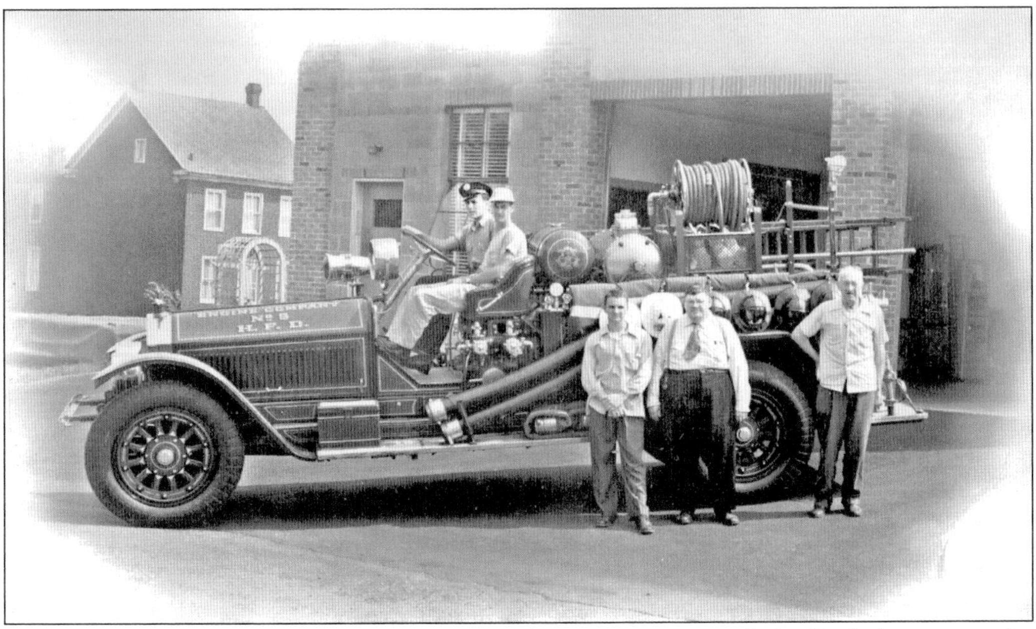

The South Hagerstown Fire Company, located at First Street and Guilford Avenue, is the youngest fire company in the city. The company was formed *c.* 1932. The social hall was known as Guilford Gardens. The firehouse was not completed until 1950. This photograph was taken on June 24, 1950, shortly after the company went into service. From left to right are Driver William Dieterich, Guy Boward, Ted Cousins, Herman Stouffer, and ? Lambert. (Courtesy of the South Hagerstown Fire Company.)

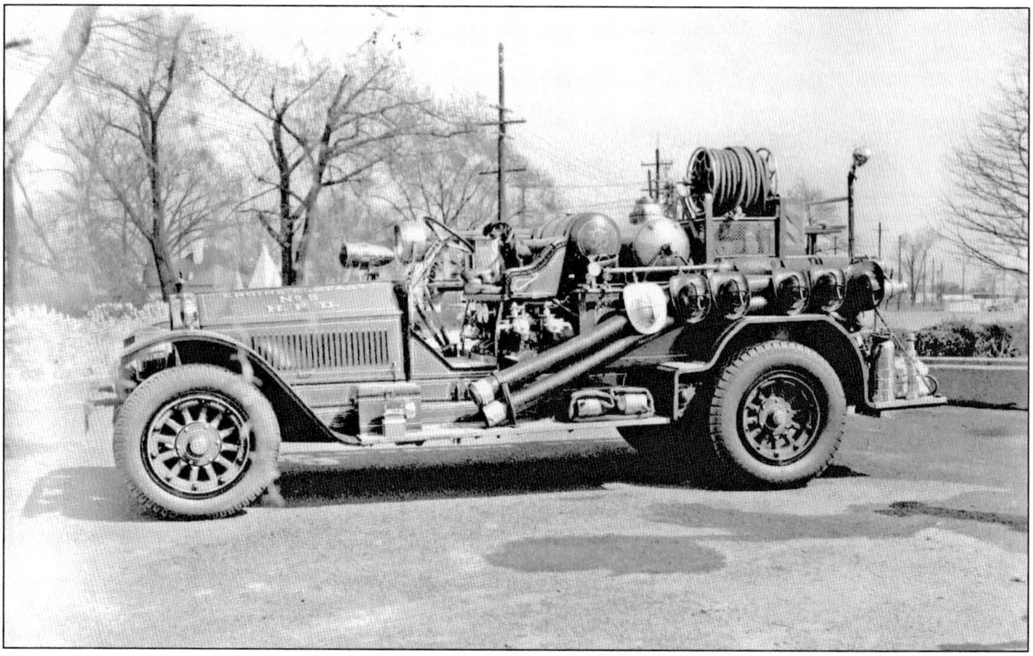

This 1925 American LaFrance pumper was used by the Western Enterprise Fire Company from 1925 until 1950. This became South Hagerstown's first pumper in 1950. It has a 750-gallon-per-minute rotary gear pump. The black dog in the picture is "Midnight." (Courtesy of the William Dieterich collection.)

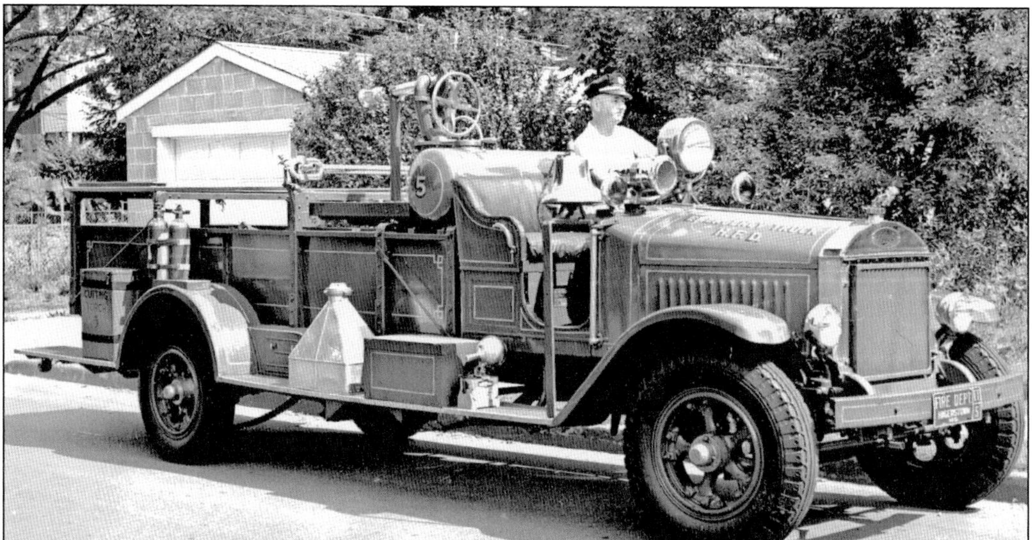

Second-generation driver Charles Flook is sitting behind the wheel of Emergency Truck 5 (ET-5), nicknamed "Snorky." Driver Flook began his career on January 1, 1953. He served in excess of 30 years and retired on November 3, 1983. Emergency Truck 5 became a second-generation truck. It first served the Pioneer Hook & Ladder Company (see page 76) as the chimney chaser. At the South Hagerstown firehouse, it served as a utility-type vehicle. It carried a mechanical foam generator and extra equipment, and it had a deluge gun mounted on top. (Courtesy of the C. William Karn collection.)

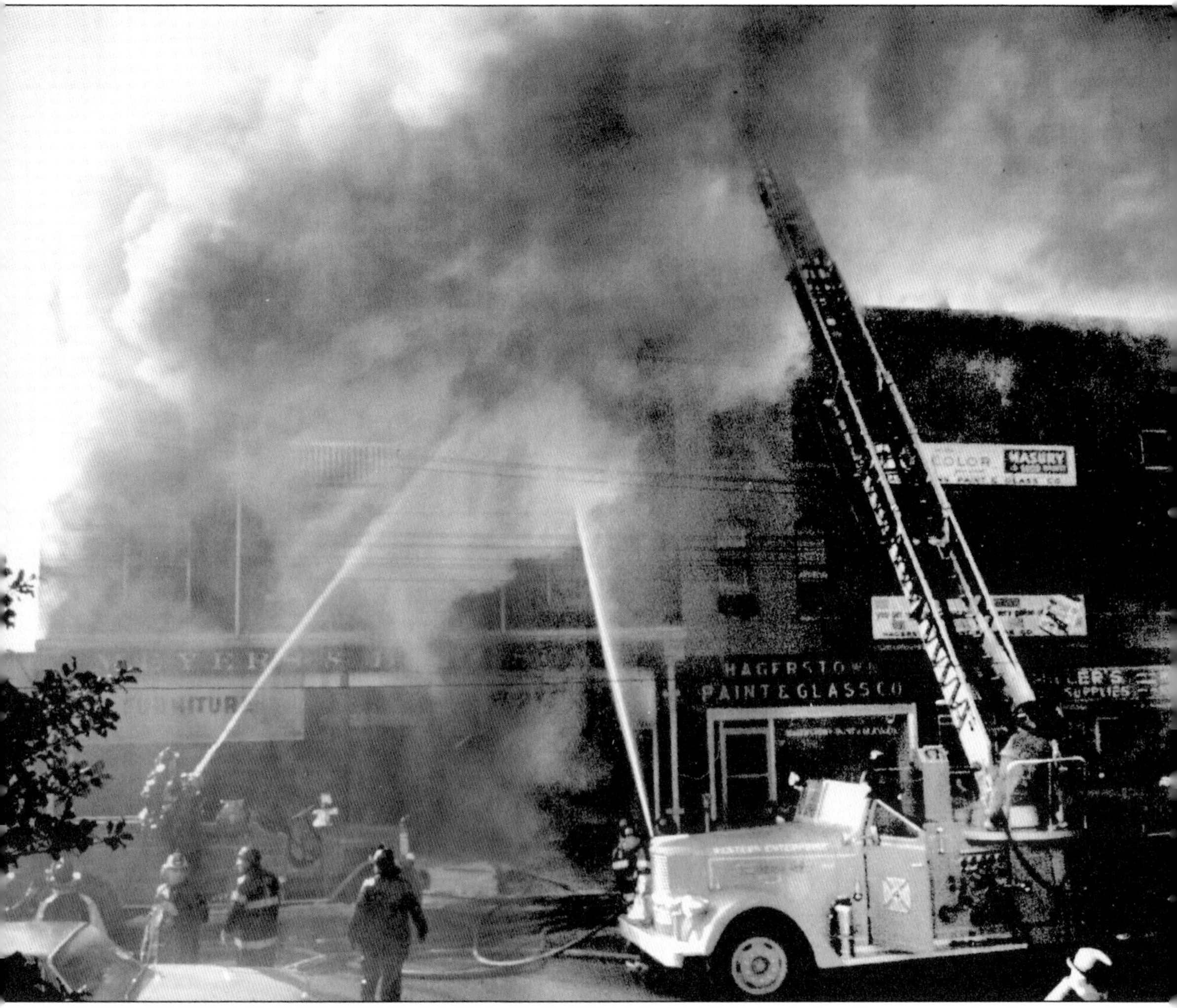

Emergency Truck 5 is pouring thousands of gallons of water onto the flames at the Myers-Berkson fire in the unit block of West Franklin Street on January 9, 1963. Emergency Truck 5 was very versatile and would be placed close enough to buildings for the fire stream to be most effective. It was placed in dangerous positions in place of a pumper. Western Enterprise Truck 4 ("Bruno") with its aerial extended is on the right. Numerous citizens were evacuated just before the smoke and flames consumed the building. (Courtesy of the First Hagerstown Hose Company.)

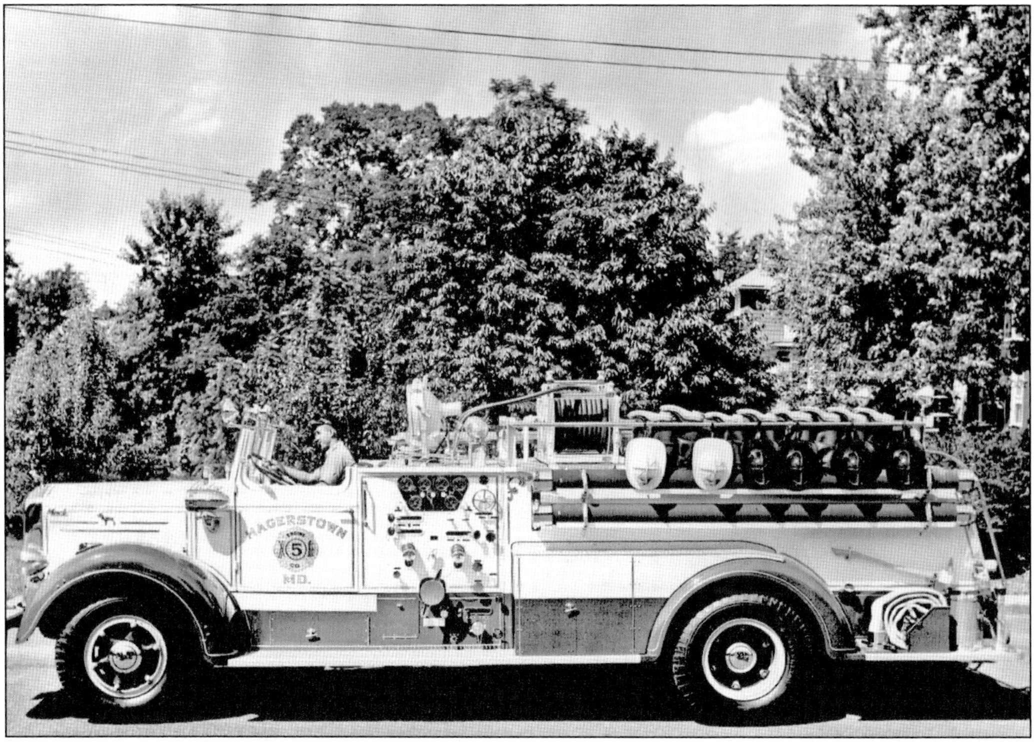

Here is driver Richard Trovinger behind the wheel of South Hagerstown's first new piece of apparatus, a 1952 Mack pumper. It has a 750-gallon-per-minute pump and a 300-gallon water tank, and cost $21,000. It bears the nickname "Bulldog." This photograph was taken in 1953. Note the large flood lights above the pump. (Courtesy of the Washington County Historical Society.)

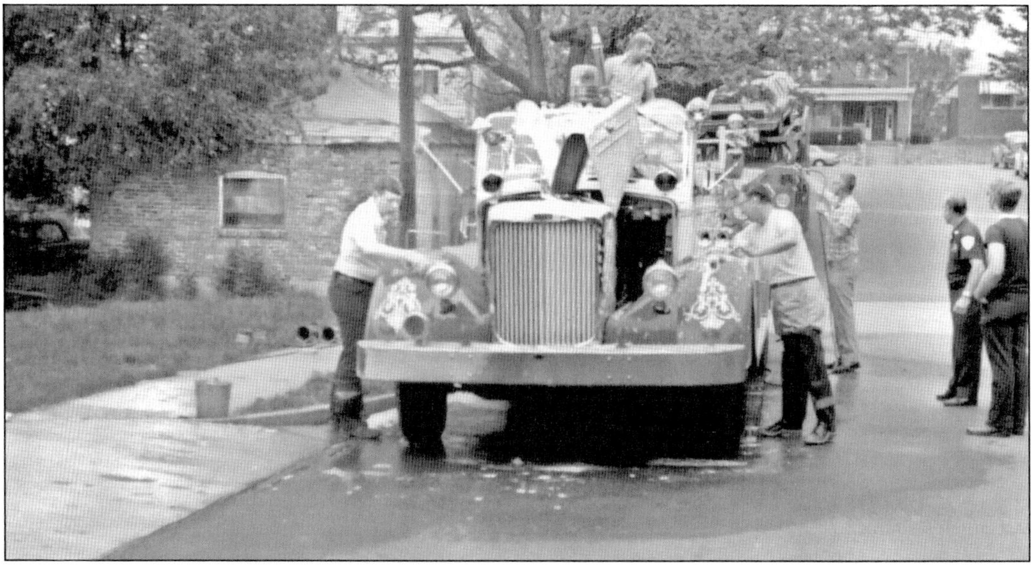

Washing fire apparatus is not performed only to keep it clean, but also to ensure the equipment is ready for service. Members of South Hagerstown are seen here cleaning the 1952 Mack in front of the firehouse. (Courtesy of the Richard Smith collection.)

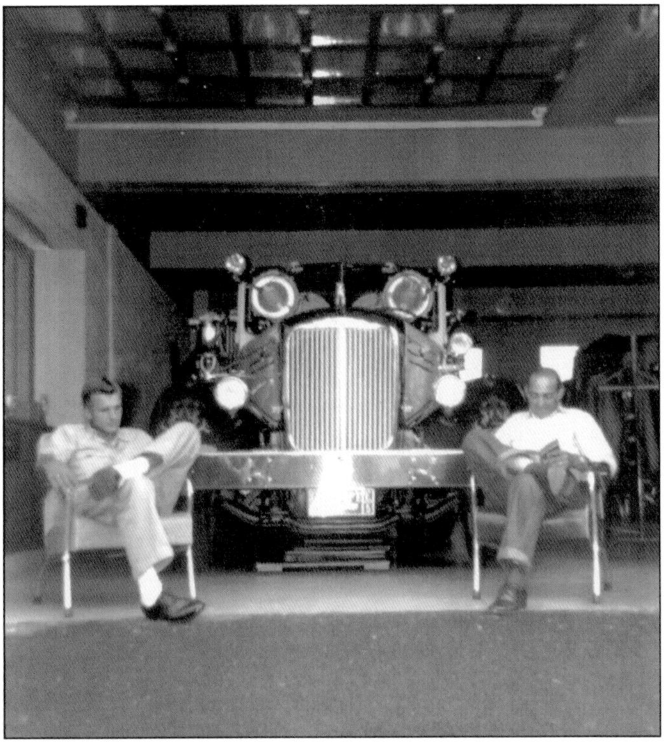

Firefighters of South Hagerstown are taking a break from the action during the summertime. Fortunately, we do not have fires every day. But firefighters never know what may happen in the next five minutes. It could be a personal injury accident (PIA) with multiple injuries or a working fire with people hanging out the windows. (Courtesy of the C. William Karn collection.)

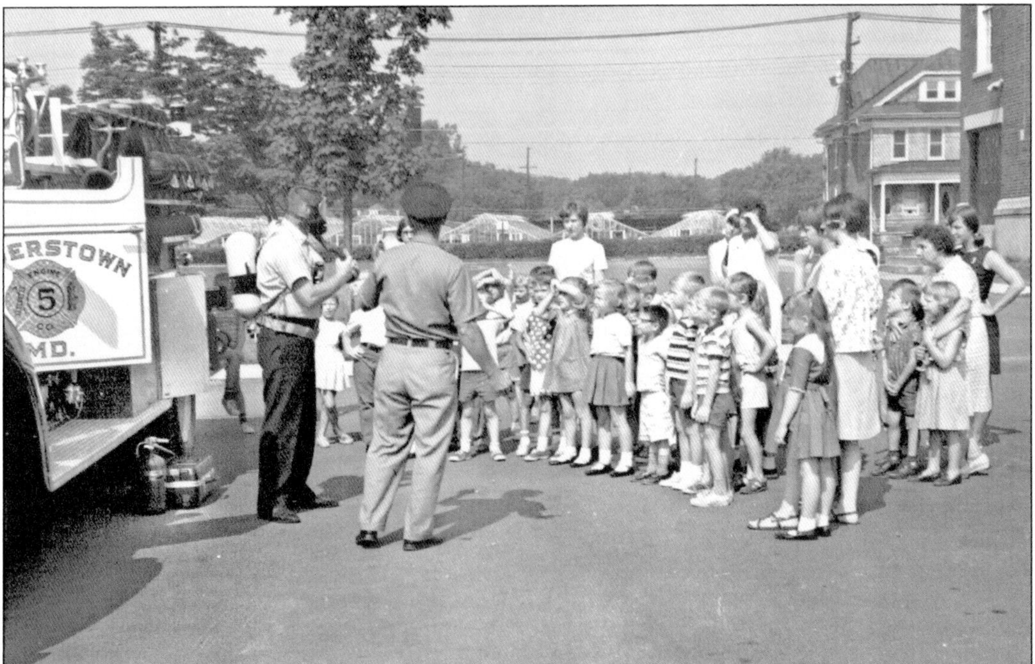

Here are South Hagerstown firefighters Charles Baker and Richard Trovinger teaching a fire-safety class to students at Bester Elementary School. It is important that children are not afraid of firefighters wearing turnout gear and self-contained breathing apparatus. (Courtesy of the First Hagerstown Hose Company.)

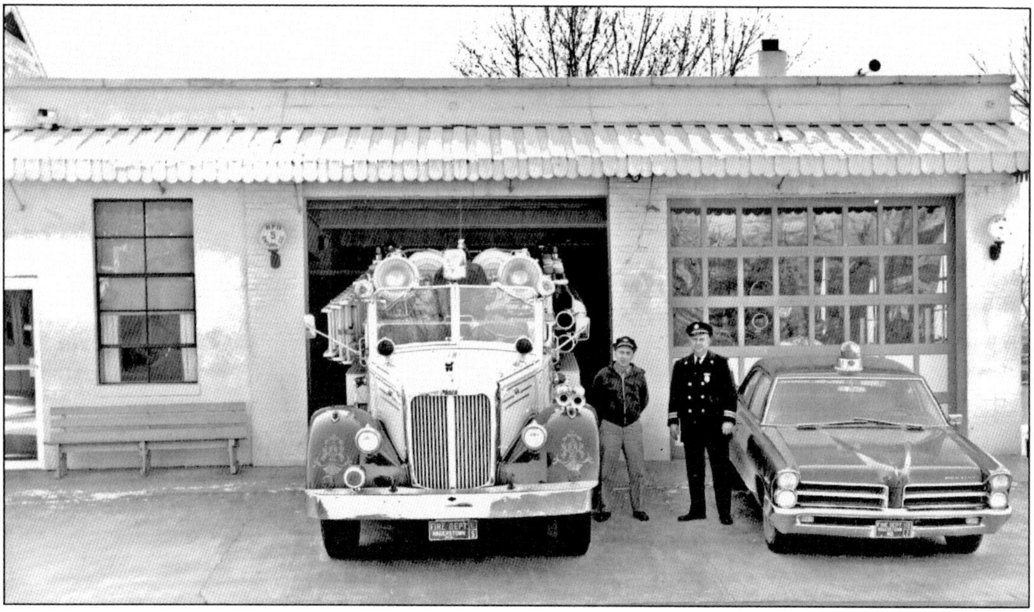

Dep. Chief C. William Karn is pictured standing with FAO Richard Trovinger. FAO Trovinger began his career with the city on May 2, 1953. After spending a brief time as a relief driver, the rest of his career was at South Hagerstown. He retired on December 29, 1983. This photograph was taken in 1972. Note the Gamewell fire alarm horn in the background. (Courtesy of the First Hagerstown Hose Company.)

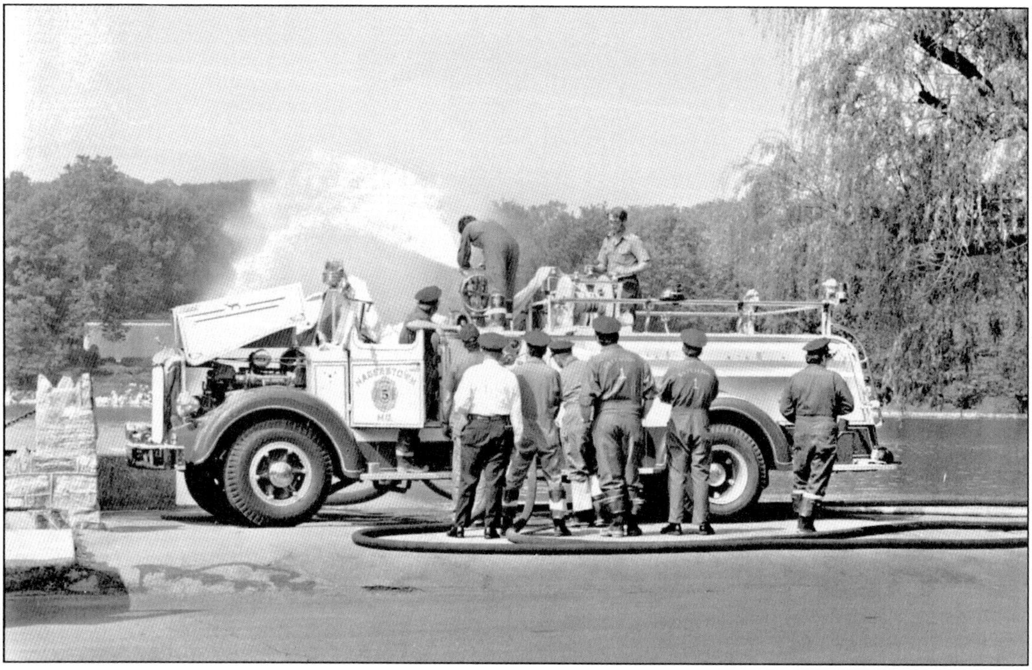

A fire apparatus operator must be proficient in driving skills, knowledge of the streets, placement of apparatus on incident scenes, hazardous materials, firefighting tactics, emergency medical services, and hydraulics, just to name a few. Firefighters are seen here pumping water with the 1952 Mack at the city park. (Courtesy of the First Hagerstown Hose Company.)

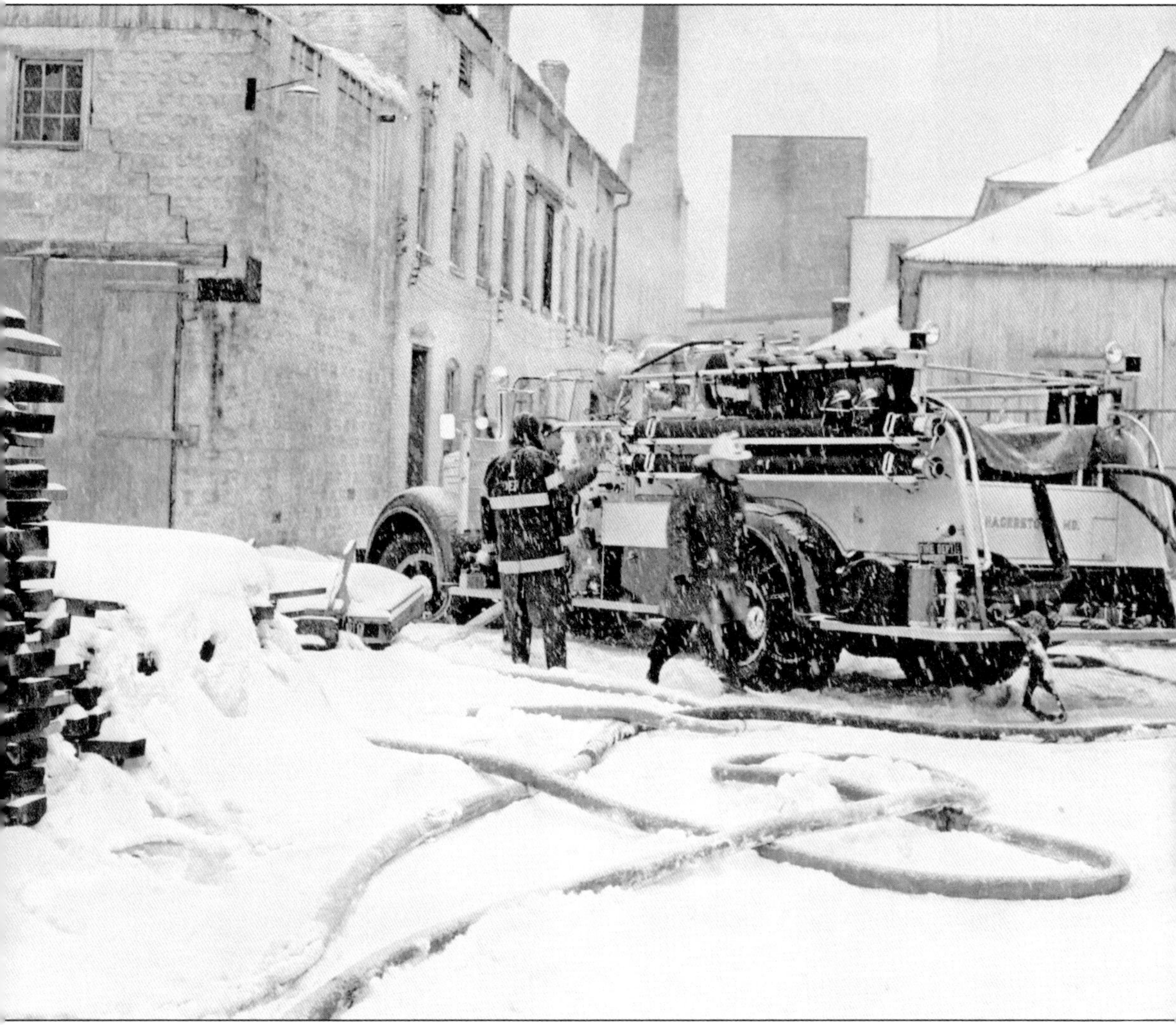

Fighting fires in the winter presents special challenges. Hose lines and pumps freeze and firefighters are chilled to the bone. These firefighters from the South Hagerstown Fire Company are seen here battling a blaze at Victor Products on Pope Avenue on January 26, 1961. Victor Products manufactured refrigerating apparatus. Prior to 1940 the building was used for manufacturing other products, including bicycles, automobiles (including Pope Tribunes, Dagmars, and taxicabs), and caskets for the Montrose Casket Company. (Courtesy of the First Hagerstown Hose Company.)

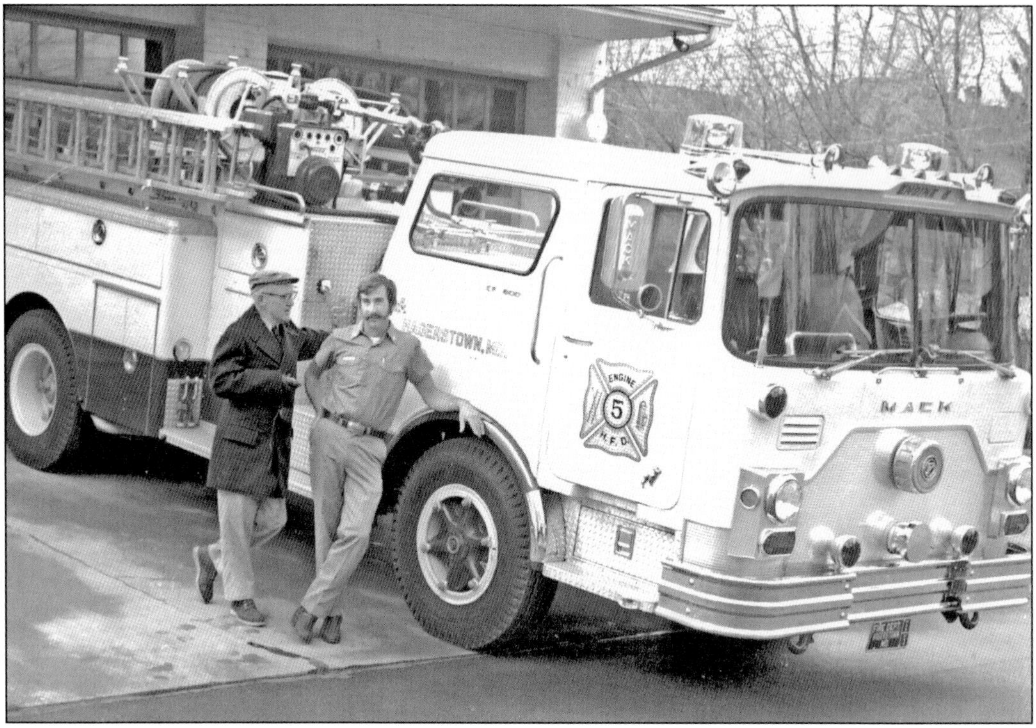

FAO Glen Gorman (right) is seen standing with retired driver A.K. McGraw in front of the South Hagerstown firehouse. The pumper is a 1972 Mack. It had a 1,000-gallon-per-minute pump and a 300-gallon water tank. This pumper went permanently out of service on February 26, 2004. (Courtesy of the Richard Smith collection.)

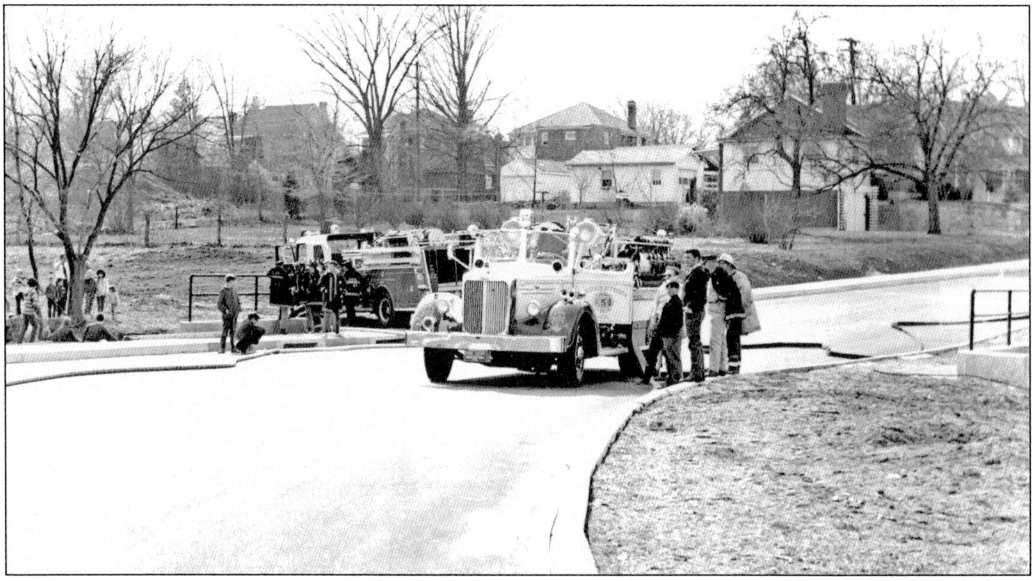

Firefighting skills must be constantly practiced. The skills learned in a training class may not be used for several years but must be remembered at a moment's notice. This is a Maryland Fire and Rescue Institute advanced training class being conducted on Magnolia Avenue in 1971. (Courtesy of the Rick Kipe collection.)

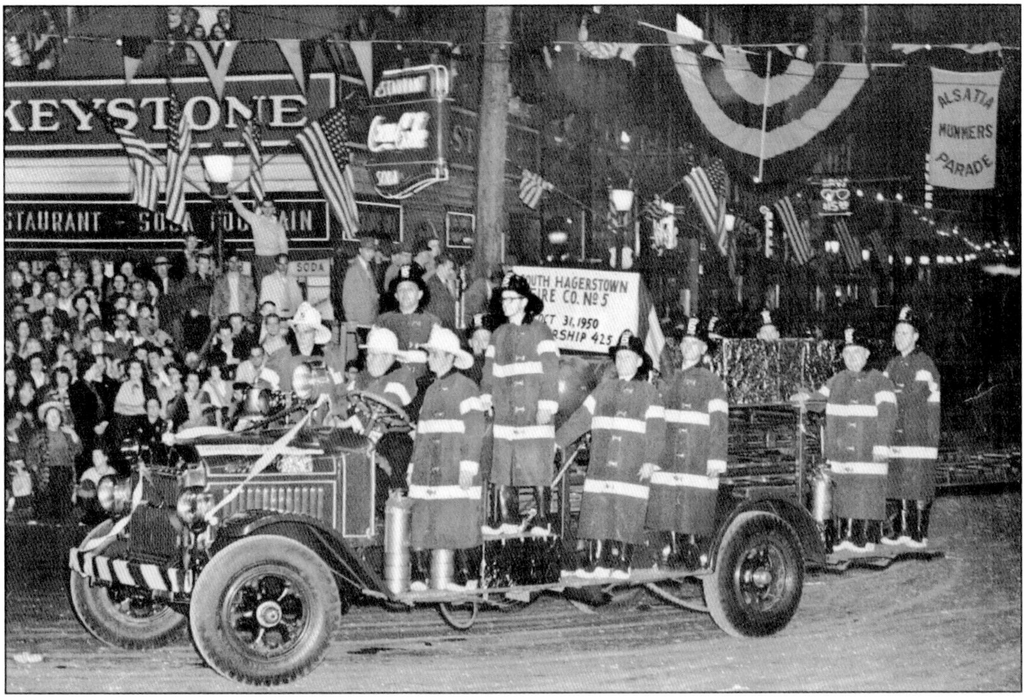

This 1927 Mack is carrying a full regiment of firefighters during the Alsatia Mummers parade. From left to right, wearing the white officers helmets, are Teddy Cousins, Herman Stouffer, and Bill Andrews. On the right side of the apparatus are Bud Lowman and Paul Potteroff. On the left side of the apparatus are Dick Shilling, Frank Day, Vic Springer, Walter Haynes, and Ralph Jacobs. (Courtesy of the South Hagerstown Fire Company.)

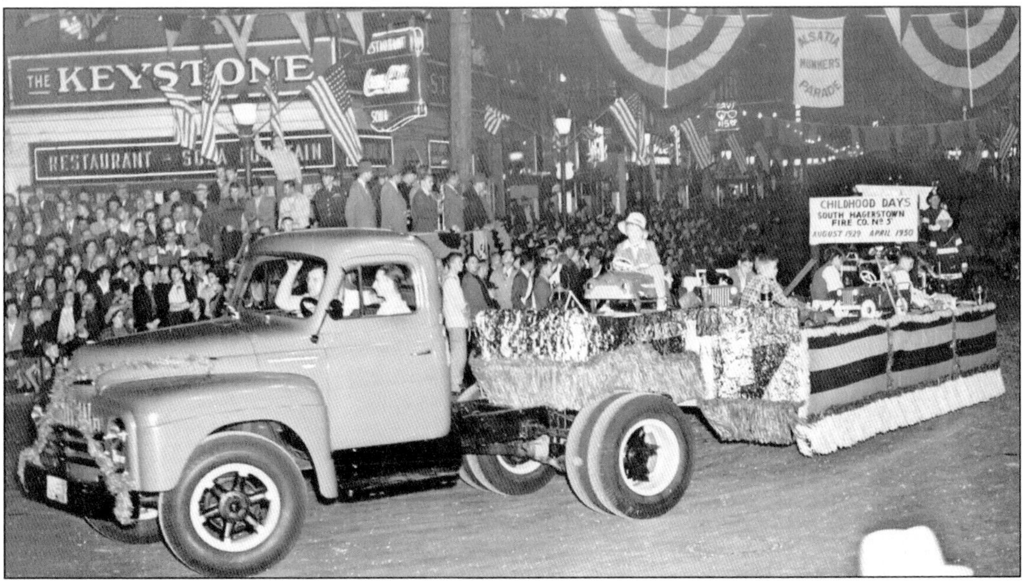

Large crowds have always lined the streets of the Alsatia Mummers parade. This crowd is watching the South Hagerstown float pass by in October 1950. The driver is Rosie Miller and the passenger is Eddie Remsburg. The photograph was taken in the public square. (Courtesy of the South Hagerstown Fire Company.)

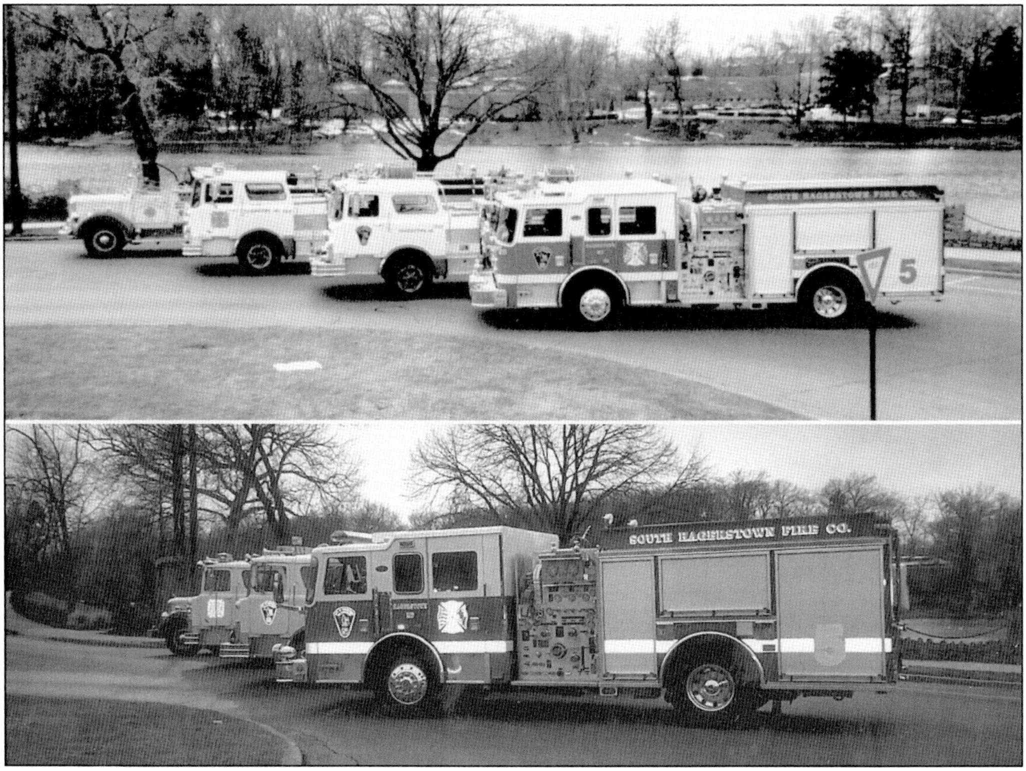

Displayed here are four different pumpers that served in the south end of Hagerstown for six decades. Shown from left to right are the 1952 Mack, 1972 Mack, 1982 Mack, and 2004 KME. The 2004 KME was placed into service on the day of this photograph, February 26, 2004. In the background is the city park. (Courtesy of the author.)

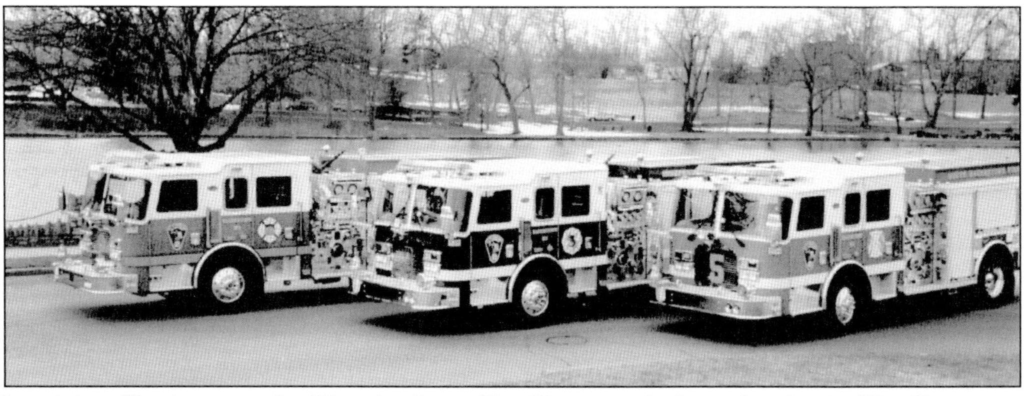

In a joint effort between the City, Antietam Fire Company, Independent Junior Fire Company, and the South Hagerstown Fire Company, these three pumpers were purchased. The three KME pumpers replace three 1982 Mack pumpers. The new pumpers have 1,500-gallon-per-minute pumps and 500-gallon water tanks. (Courtesy of the author.)

Six

Pioneer Hook & Ladder Company, Truck 1

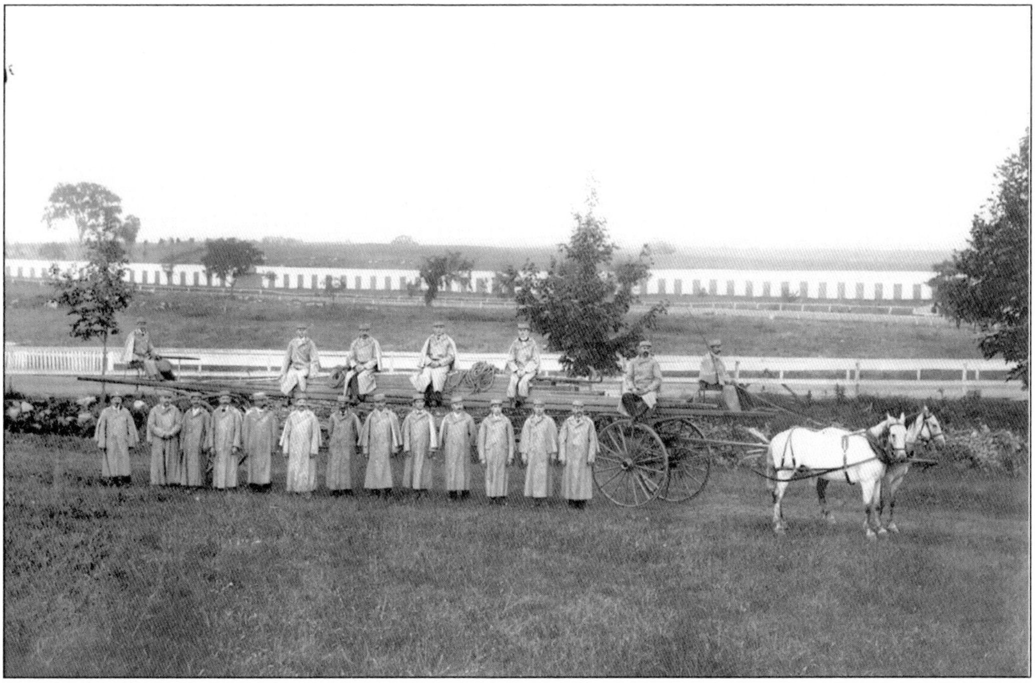

Members of the Pioneer Hook & Ladder Company are shown standing with their first ladder truck. The photograph was taken at the Hagerstown fairgrounds. The first truck was delivered on October 17, 1872. Note the horse-race track in the background. (Courtesy of the Pioneer Hook & Ladder Company.)

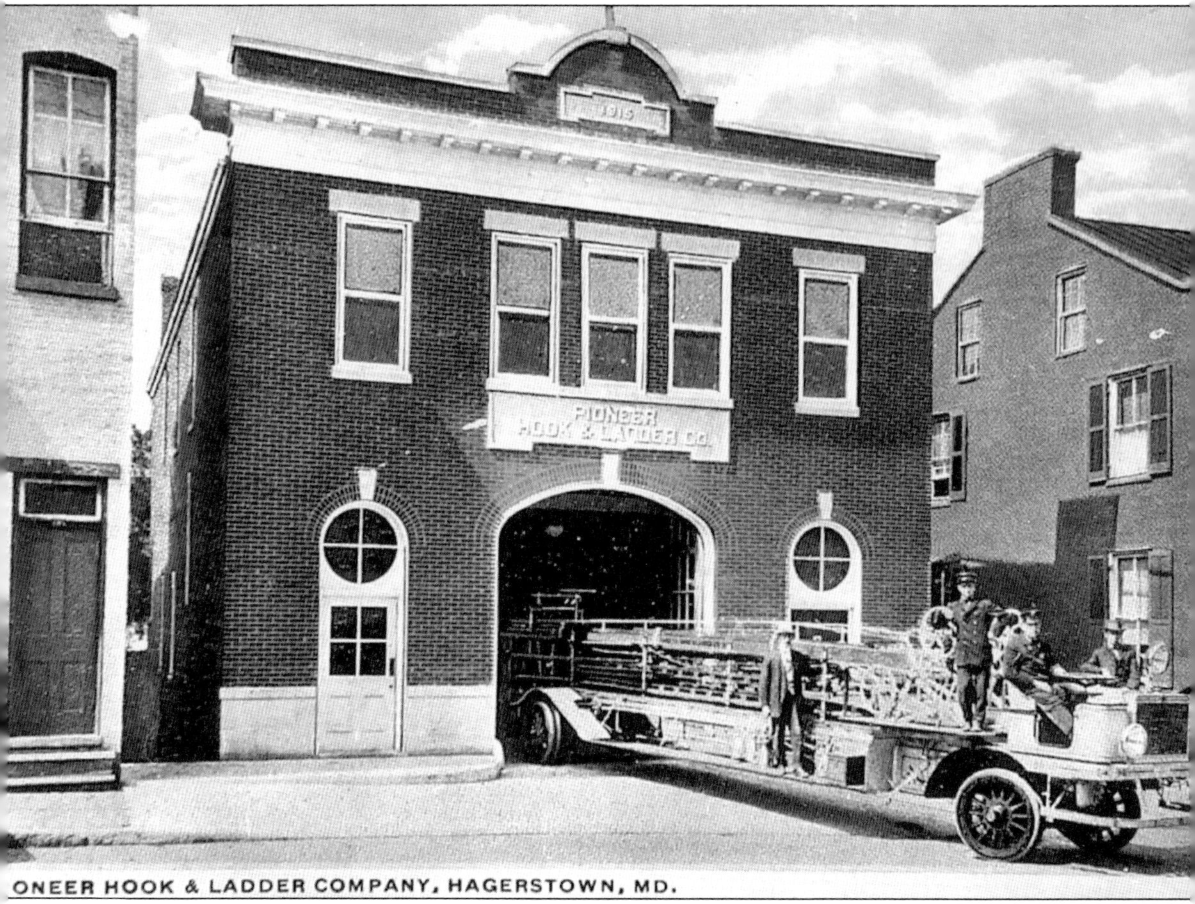

After the burning of the courthouse, firefighter William H. Armstrong (who served as president of the company for 50 years) convinced the mayor and council of the need for a hook-and-ladder truck. An appropriation of $800 was made to purchase the apparatus. The Pioneers were organized on August 15, 1872. The apparatus was kept at the Market House. By 1894 the hook-and-ladder truck was housed at the Independent Junior Fire Company, but on March 4, 1914, a lease was signed with the Antietam Fire Company. After the Antietam firehouse burned on October 28, 1914, the Pioneers constructed this firehouse at 21 West Franklin Street, on the site of the Oak Spring lot. The cost of construction in 1915 was $6,000. (Courtesy of the David White collection.)

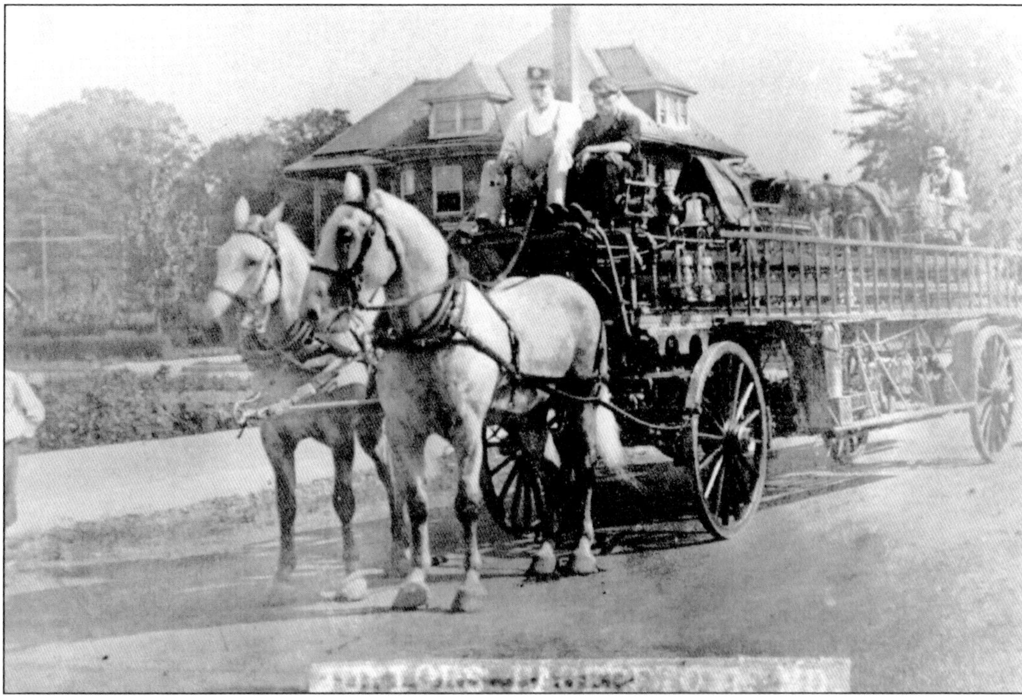

The Pioneers always had a tiller truck (a driver or tillerman steers the rear wheels) because of the length of a ladder truck in relation to the narrow streets. This is most likely the Pioneers' second ladder truck. (Courtesy of the David White collection.)

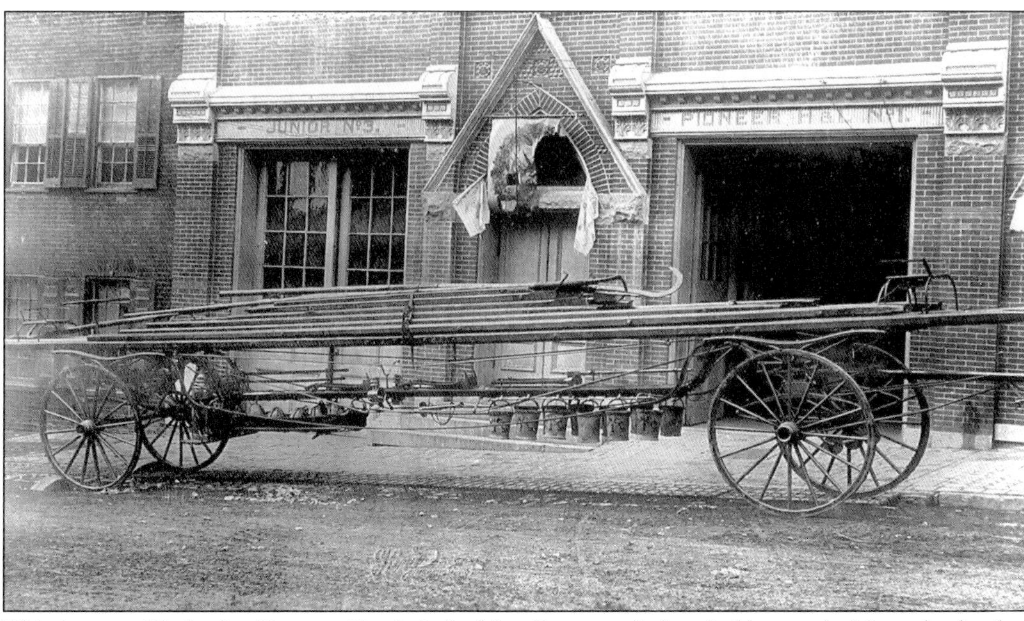

This is most likely the Pioneer Hook & Ladder Company's first ladder truck. Note the leather buckets hanging from the apparatus. This photo was taken when the Juniors shared their home with the Pioneers. (Courtesy of the Pioneer Hook & Ladder Company.)

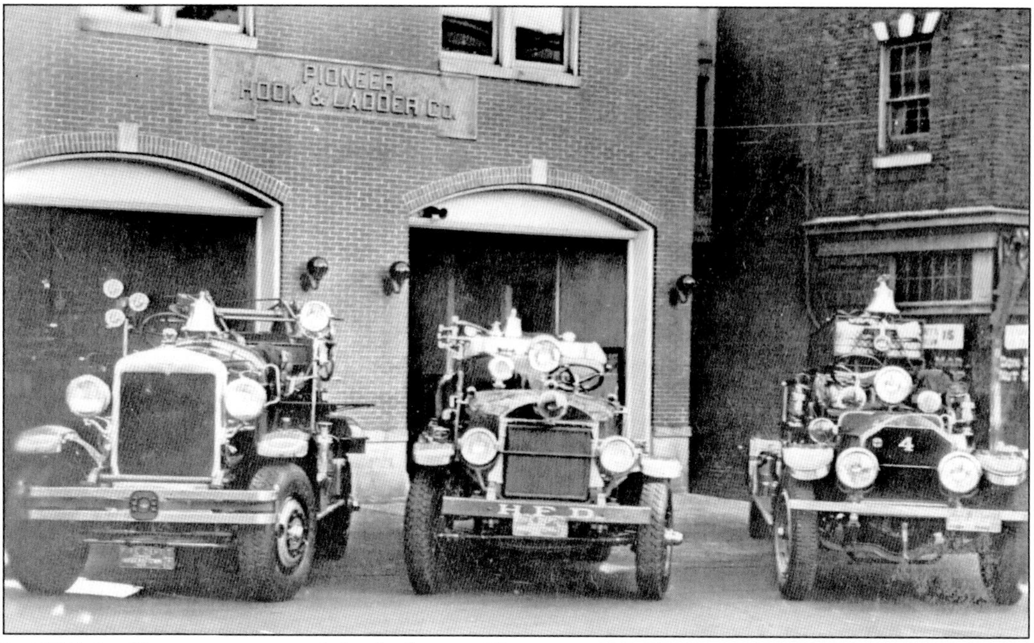

In 1934 the Pioneers renovated the firehouse. The original firehouse only had one bay. Pictured in front of the firehouse are the 1930 Ahrens-Fox ladder truck, the 1927 Mack service truck, and the Western Enterprise 1920 city service truck. (Courtesy of the William Dieterich collection.)

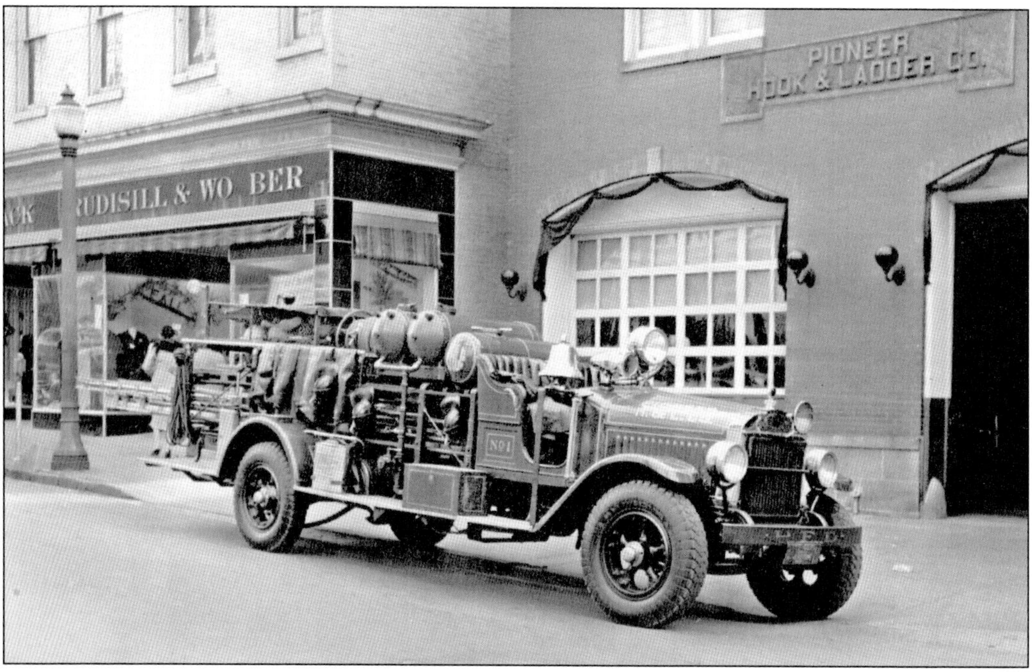

Parked in front of the firehouse on West Franklin Street is the 1927 Mack service truck known as the "Chimney Chaser." This is the truck that would later be used by the South Hagerstown Fire Company as Emergency Truck 5 (ET-5). Note the black drape hanging over the bay doors. This indicates that a member of the fire department has died. (Courtesy of the C. William Karn collection.)

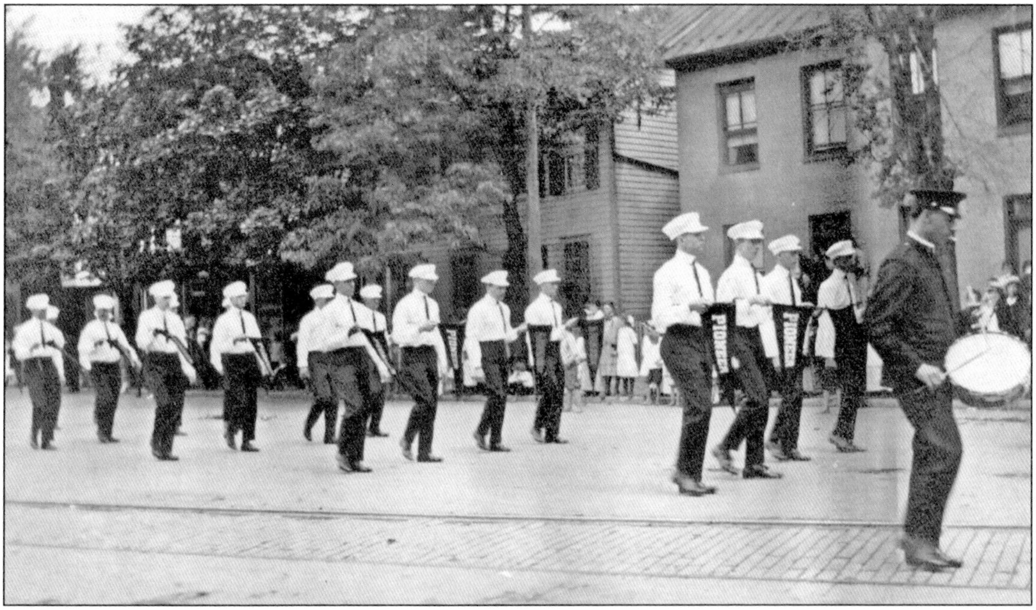

Members of the Pioneers are seen marching in a parade. They kept in step by the drum beat. Note that each member is carrying a Pioneer's pennant. (Courtesy of the First Hagerstown Hose Company.)

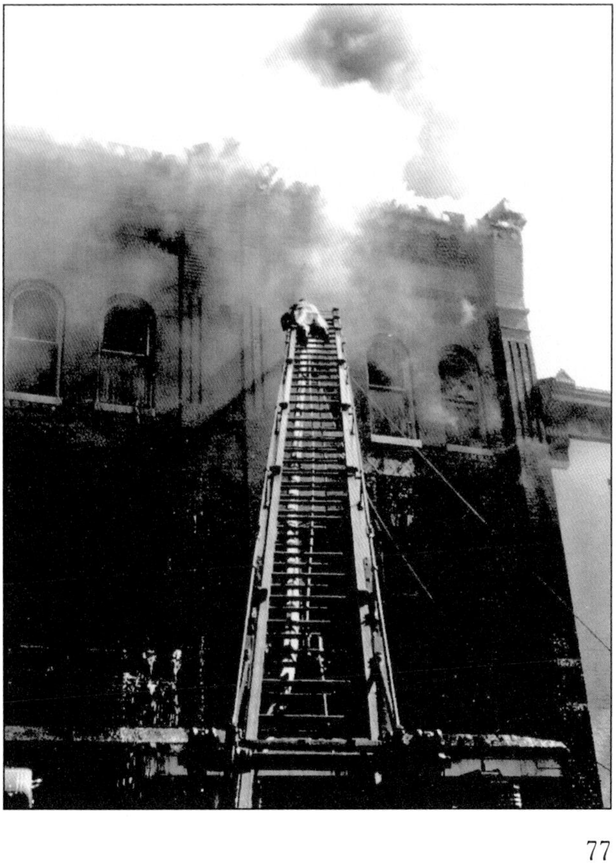

A firefighter at the tip of Pioneer's 1930 Ahrens-Fox Model 85–6–1 (Reg. No. 2049) ladder truck directs a water stream onto the S.M Bloom fire in the 100-block of North Potomac Street. (Courtesy of the William Dieterich collection.)

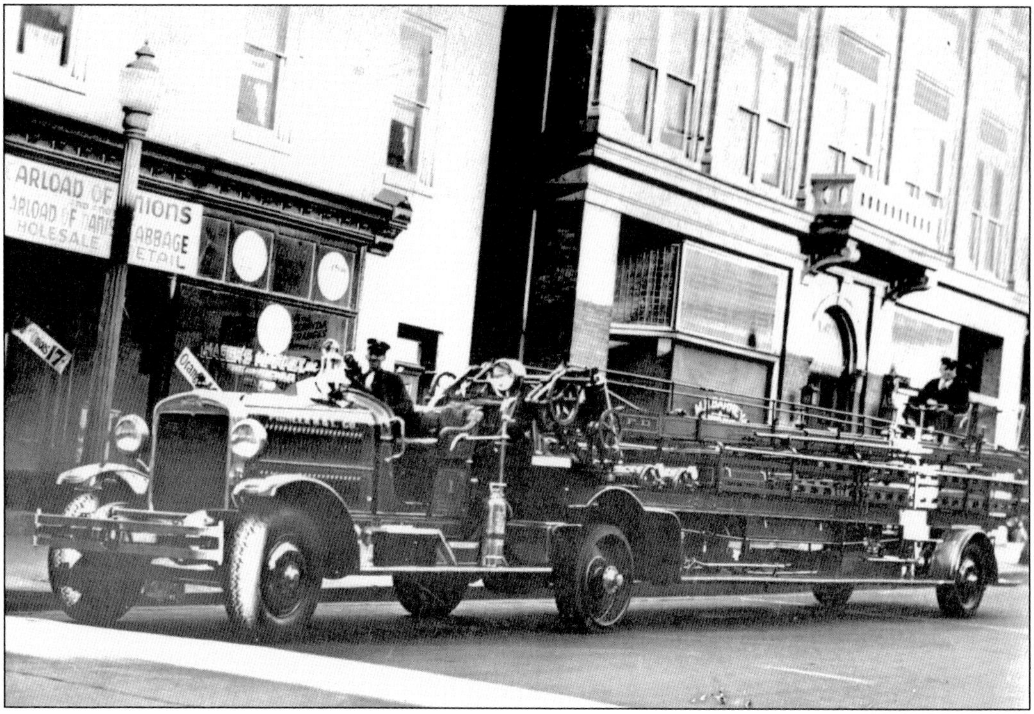

Pioneer drivers George Borne (front) and Herman Neagly (rear) are on duty with the 1930 Ahrens-Fox ladder truck known as the "85." It was called the 85 because of the length of the aerial. Driver Borne began his career as a firefighter on October 9, 1922. Before his firefighting career began, Borne was a chauffeur. (Courtesy of the William Dieterich collection.)

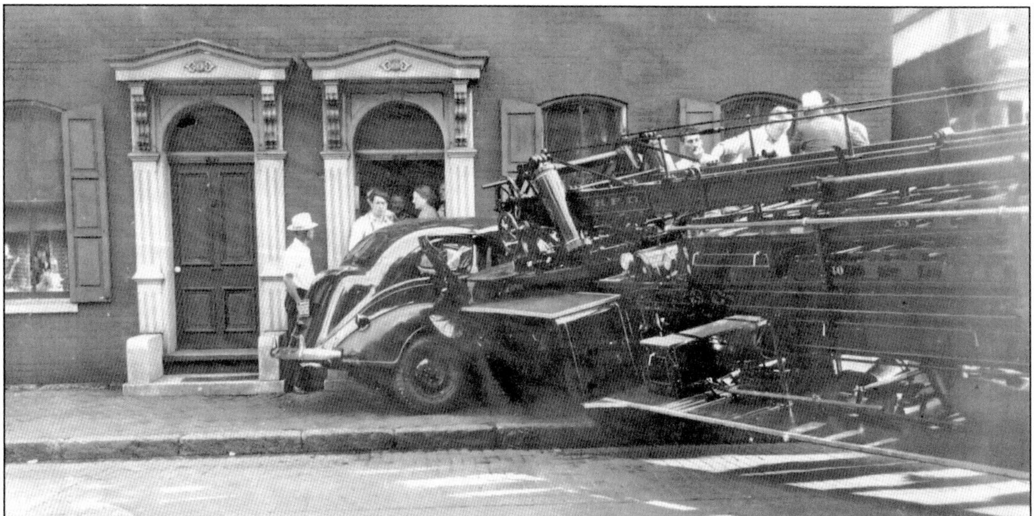

Pioneer drivers Charles Flook and William Beard were responding to a street alarm at Cannon Avenue and Baltimore Street when this accident occurred. The 1930 Ahrens-Fox ladder truck struck a car, driven by a volunteer firefighter responding to the same incident, at the intersection of East Franklin and North Locust Streets. The street alarm turned out to be a malicious false alarm; a child standing on a tricycle had pulled it. The ladder truck was sent back to the manufacturer to be refurbished. (Courtesy of the George Basore collection.)

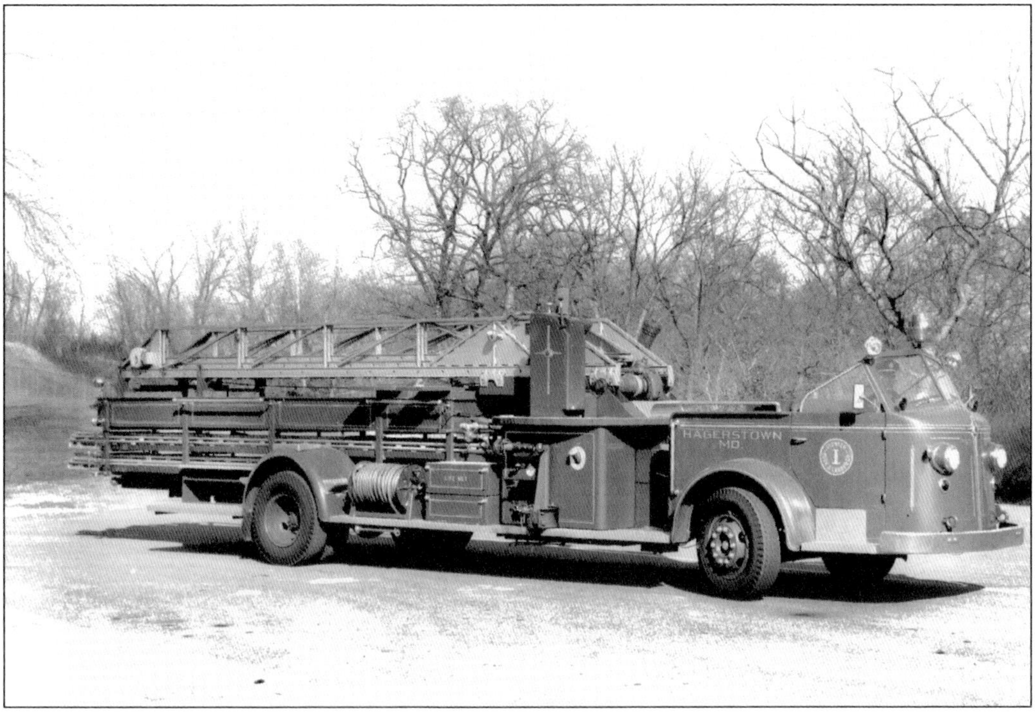

This 1948 American LaFrance ladder truck replaced the 1927 Mack service truck. This truck was known as the "65" because of the 65-foot aerial ladder. It was eventually sold to West Point, Virginia. (Courtesy of the Rick Kipe collection.)

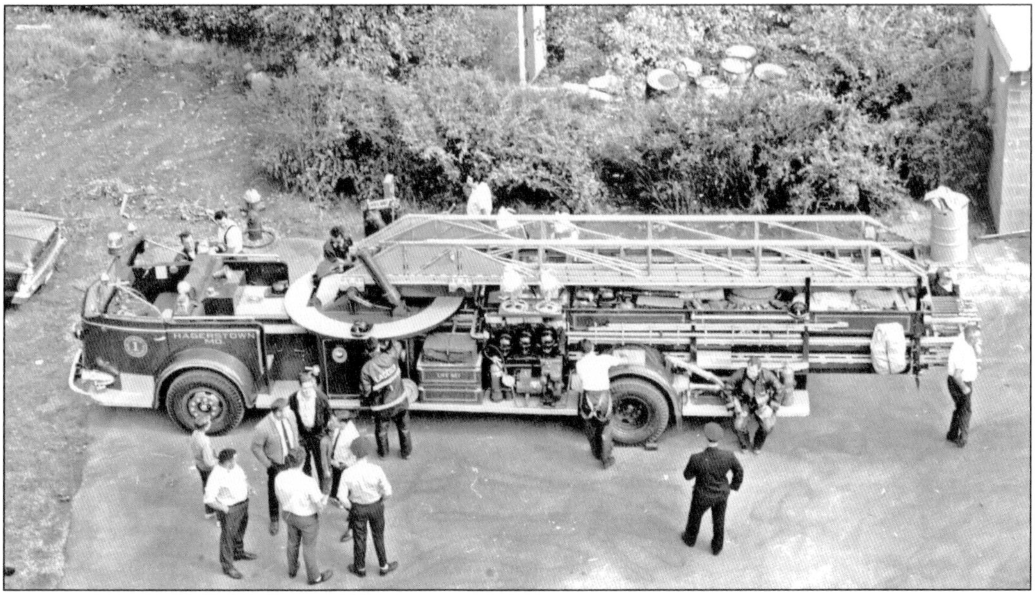

This is a bird's-eye view of the 1948 American LaFrance ladder truck. The photograph was taken from the training tower at the Hagerstown Training Center on Bowman Avenue. Most early-model fire apparatus had open cabs. The open cab permitted the driver good visibility while determining where to position the aerial during firefighting. Open cab models had windshield wipers on the inside as well as the outside. (Courtesy of the Rick Kipe collection.)

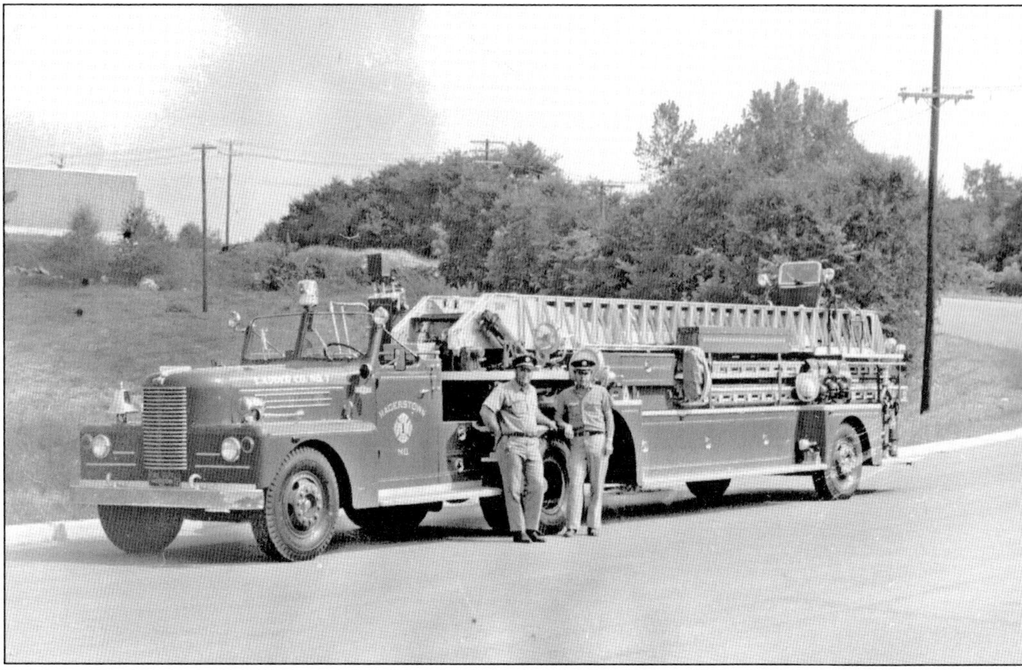

FAOs Charlie McCoy (left) and Ronald Moser pose in front of the 1959 Pirsch ladder truck. The photograph was taken on Bowman Avenue. The truck has a 100-foot aluminum aerial. (Courtesy of the First Hagerstown Hose Company.)

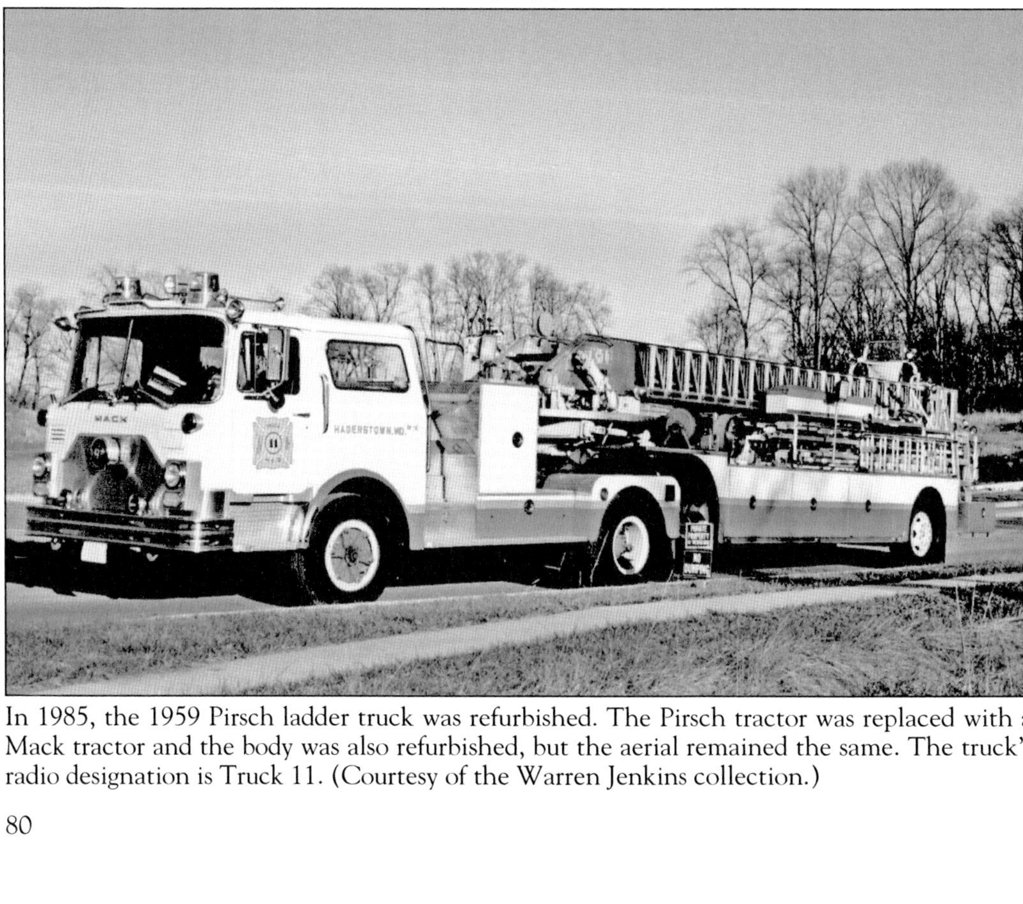

In 1985, the 1959 Pirsch ladder truck was refurbished. The Pirsch tractor was replaced with a Mack tractor and the body was also refurbished, but the aerial remained the same. The truck's radio designation is Truck 11. (Courtesy of the Warren Jenkins collection.)

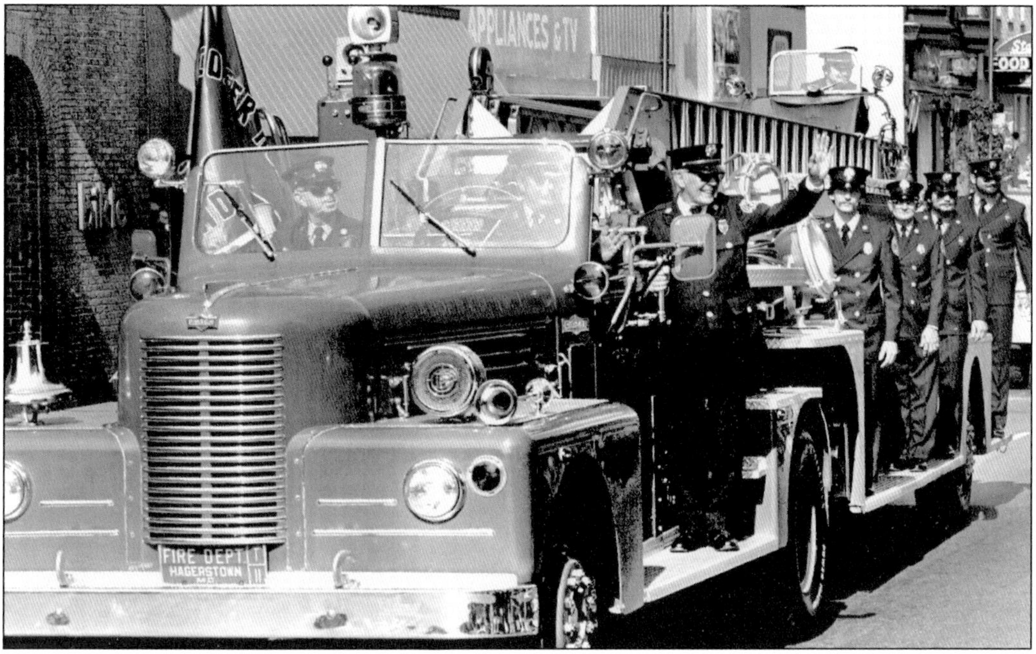

Shown here are members of the Pioneers wearing class "A" uniforms in a parade on North Potomac Street. From left to right are Charles Daley (president), Lenny Barton (driver), Capt. William Dieterich, Ronald Horn (driver/tillerman), Todd Gelwicks, Jack Emory, Dean Baker, and Doug Spickler. (Courtesy of the Todd Gelwicks collection.)

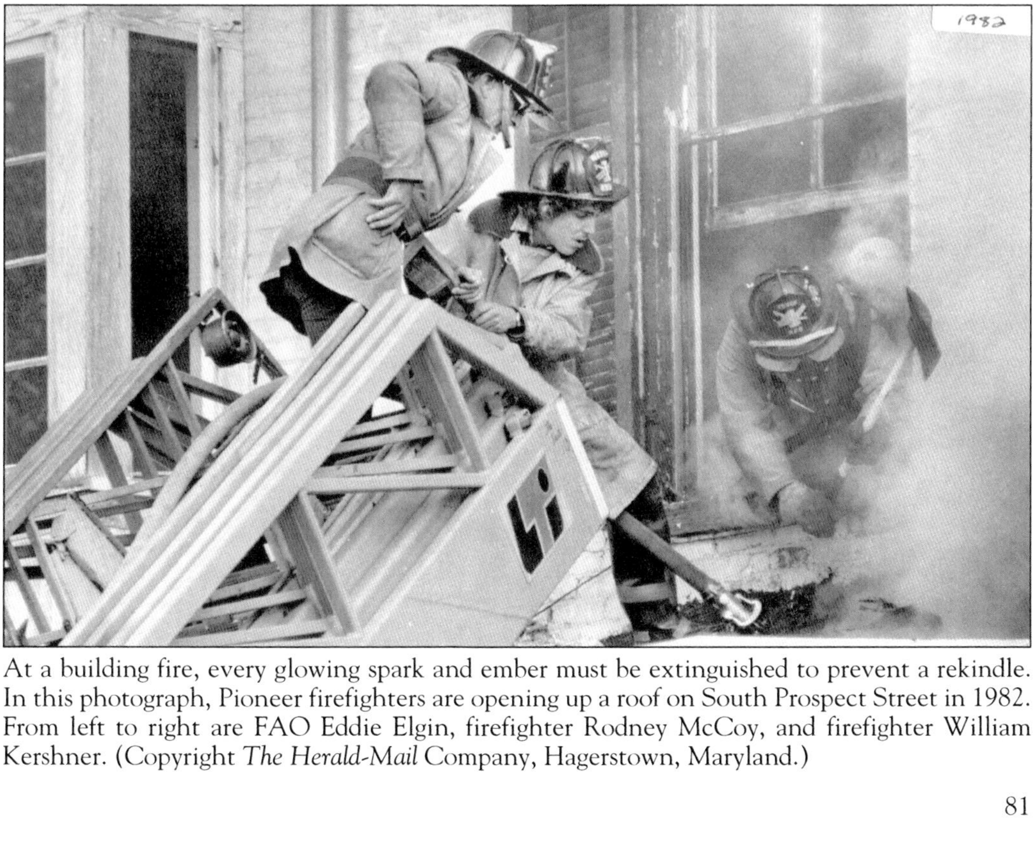

At a building fire, every glowing spark and ember must be extinguished to prevent a rekindle. In this photograph, Pioneer firefighters are opening up a roof on South Prospect Street in 1982. From left to right are FAO Eddie Elgin, firefighter Rodney McCoy, and firefighter William Kershner. (Copyright *The Herald-Mail* Company, Hagerstown, Maryland.)

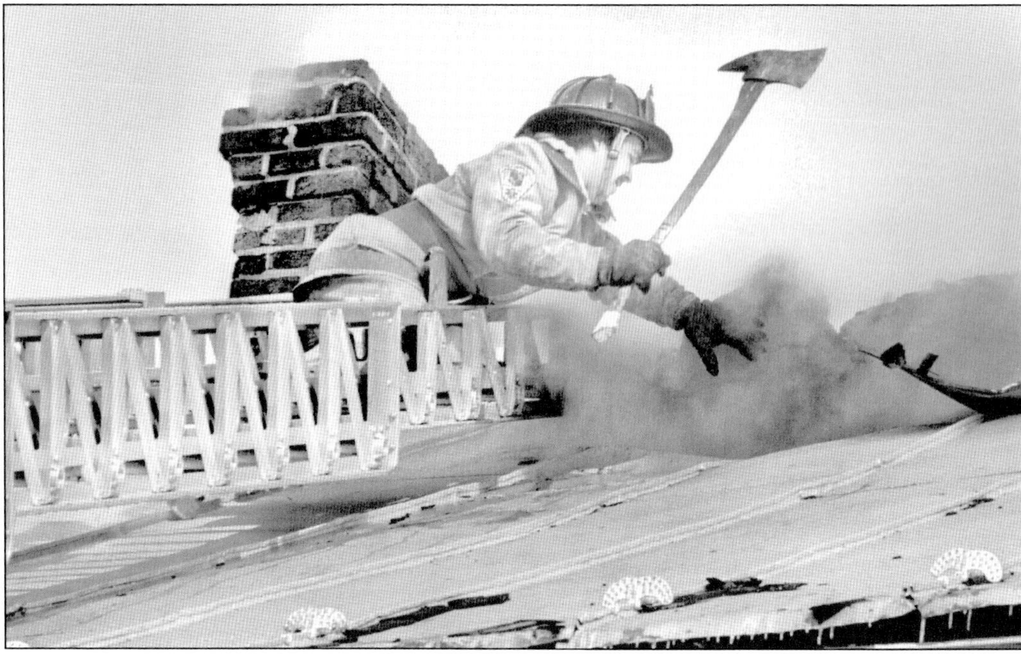

Pioneer firefighter Tim McCoy is shown ventilating a roof on East Antietam Street in 1983. The roof is opened up to remove the smoke, heat, and toxic gases from a structure. This allows firefighters the opportunity to perform search-and-rescue and fire-extinguishing maneuvers more efficiently. (Copyright *The Herald-Mail* Company, Hagerstown, Maryland.)

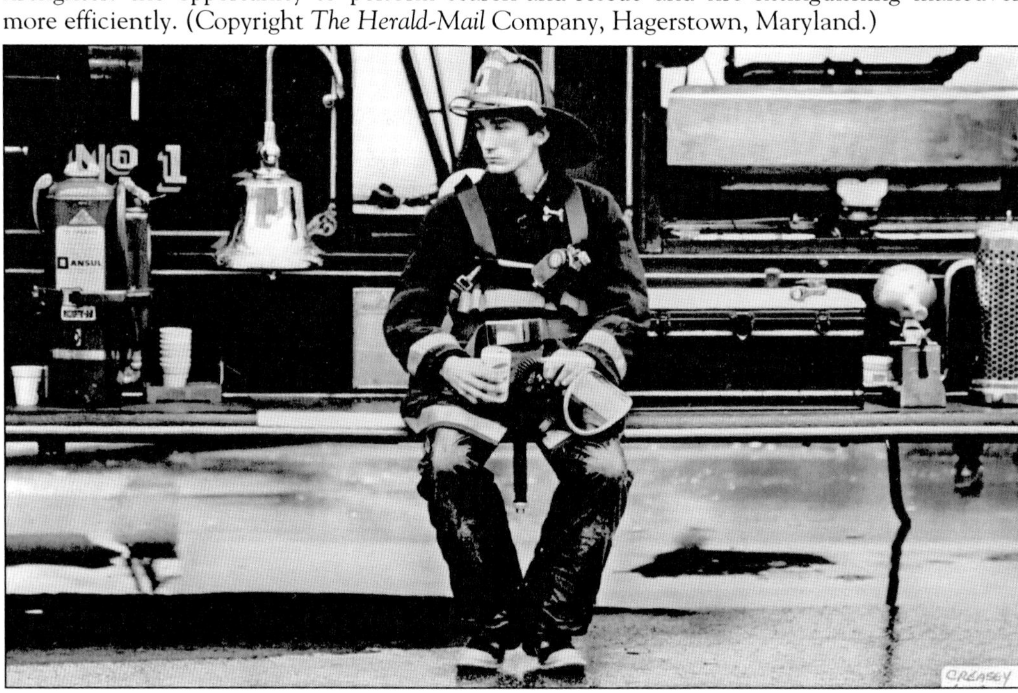

Todd Gelwicks, a firefighter at the First Hagerstown Hose Company and later at the Pioneers, takes a break on the hose company's 1946 Ahrens-Fox ("Slugger") at the Creasey Company fire on West Franklin Street on November 21, 1977. This would be Slugger's last major fire. (Copyright *The Herald-Mail* Company, Hagerstown, Maryland.)

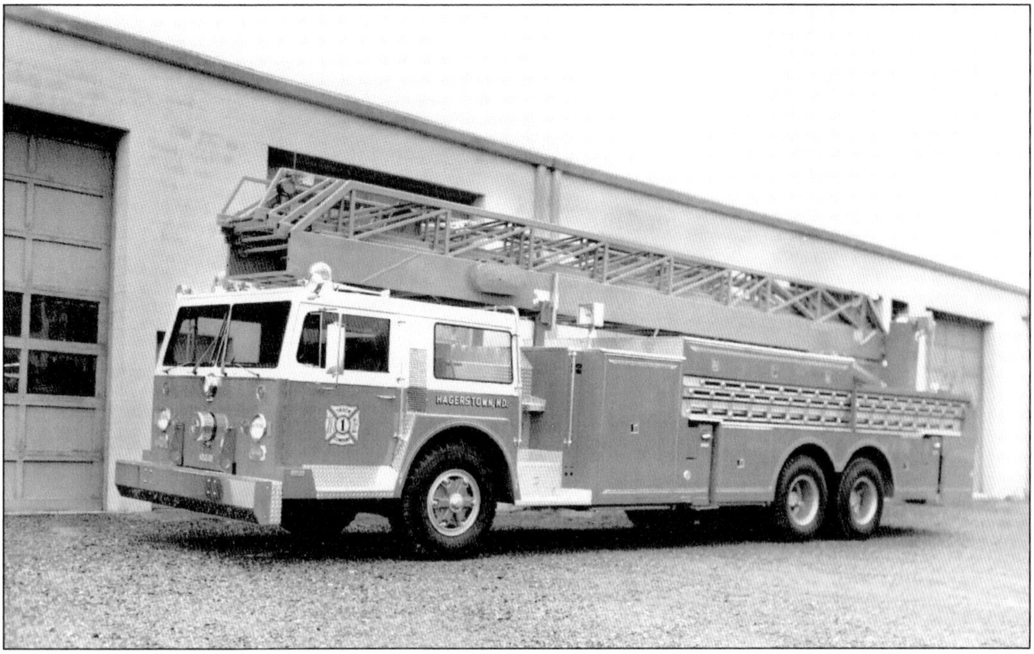

In 1973 the City purchased this Oren/Hendrickson 100-foot aerial ladder truck. This photograph was taken when the truck was brand new. Note the aerial has not been lettered yet. The aerial was built by the Grove Manufacturing Company of Shady Grove, Pennsylvania. This version of the truck also had a small booster pump. (Courtesy of the C. William Karn collection.)

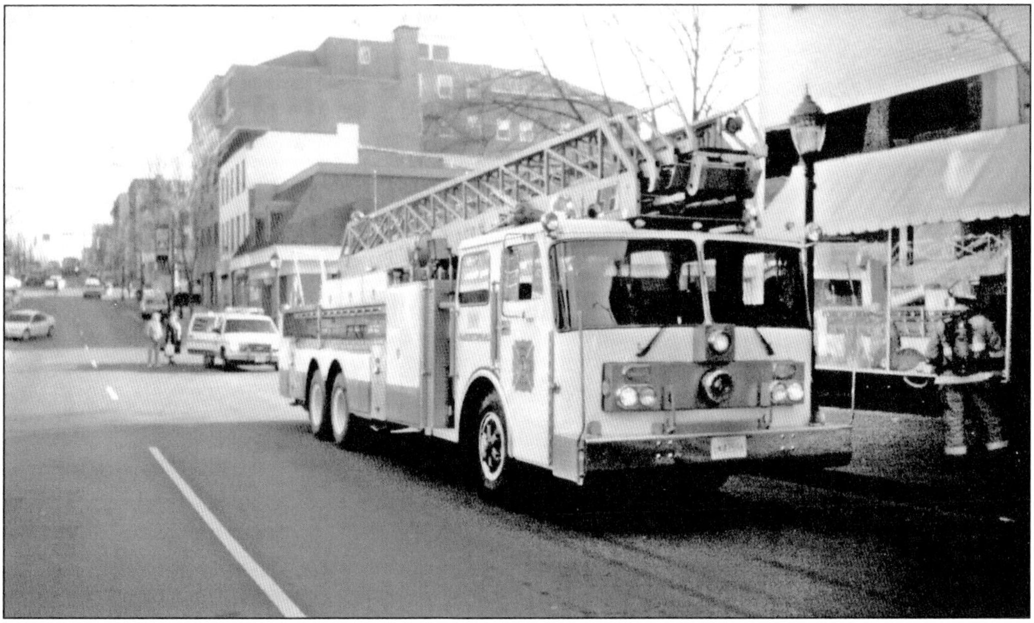

In the mid-1980s the Oren/Hendrickson was refurbished with a Spartan cab and a new paint job. The truck is known as "Brute." This photograph was taken on North Potomac Street, in front of the Elizabeth Center in 1997. This was one of Brute's last responses in Hagerstown. The truck was eventually sold to the Waynesboro, Pennsylvania Fire Department. (Courtesy of the author.)

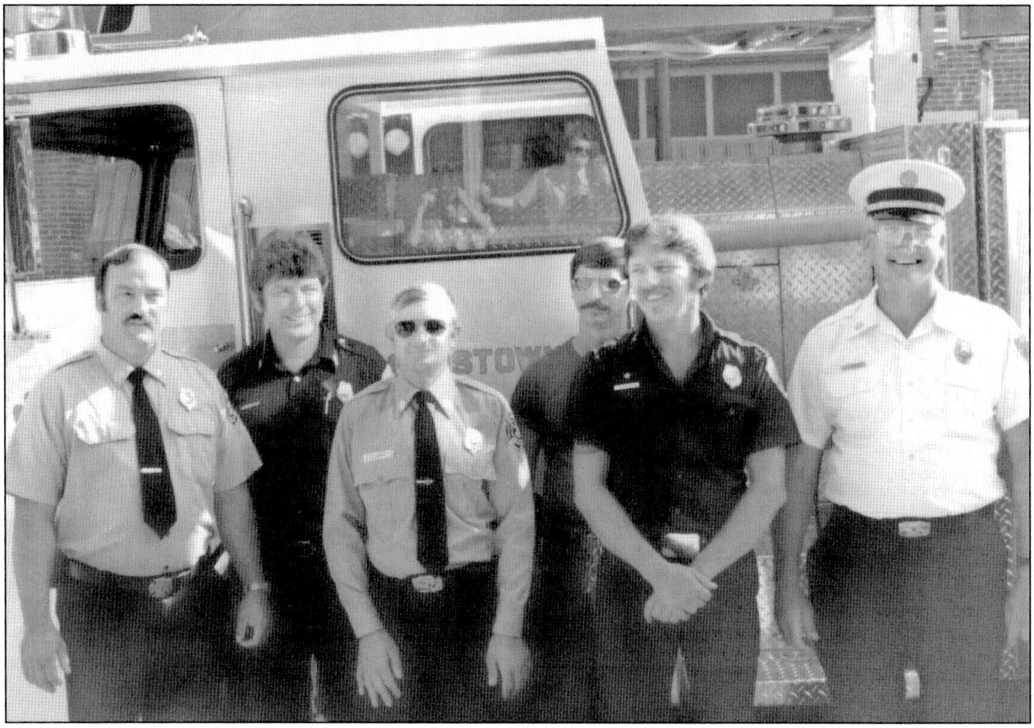

Members of the Pioneer Hook & Ladder Company and the "C" shift are standing in front of Brute. From left to right are James Winters, FAO Larry Williams, Jack Emory, Glen Giffin, FAO Andrew Hartman, and Batt. Chief Charles Baker. (Courtesy of the Richard Smith collection.)

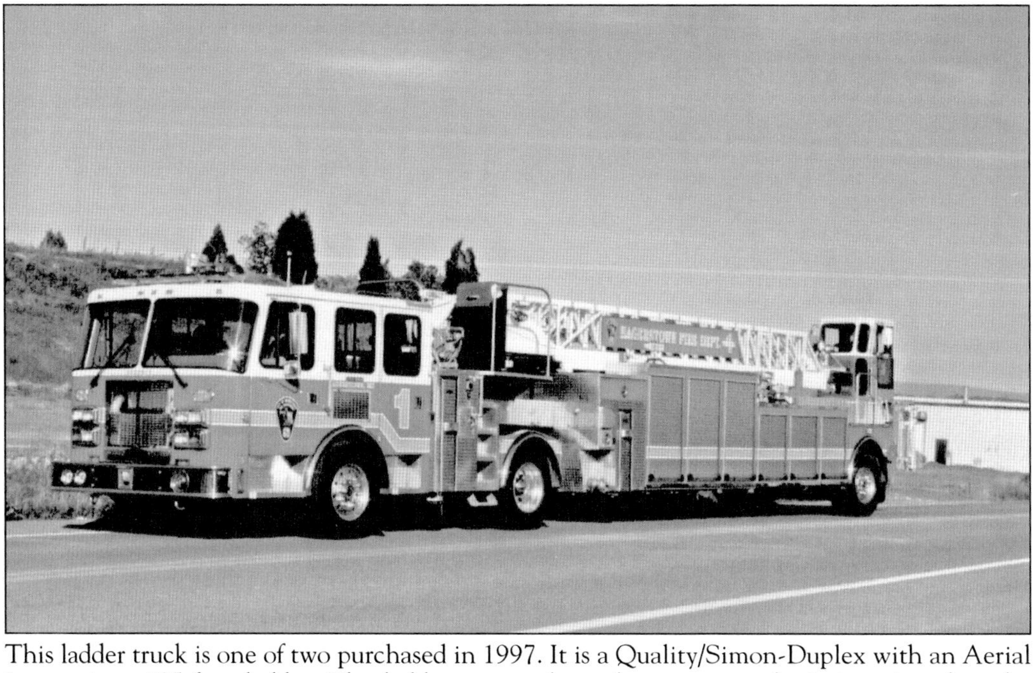

This ladder truck is one of two purchased in 1997. It is a Quality/Simon-Duplex with an Aerial Innovations 105-foot ladder. The ladder can work at almost any angle. It is assigned to the Pioneers and is designated Truck 1. (Courtesy of the David Bowen collection.)

Seven

Career Officers, Drivers, and Fire Apparatus Officers (FAO)

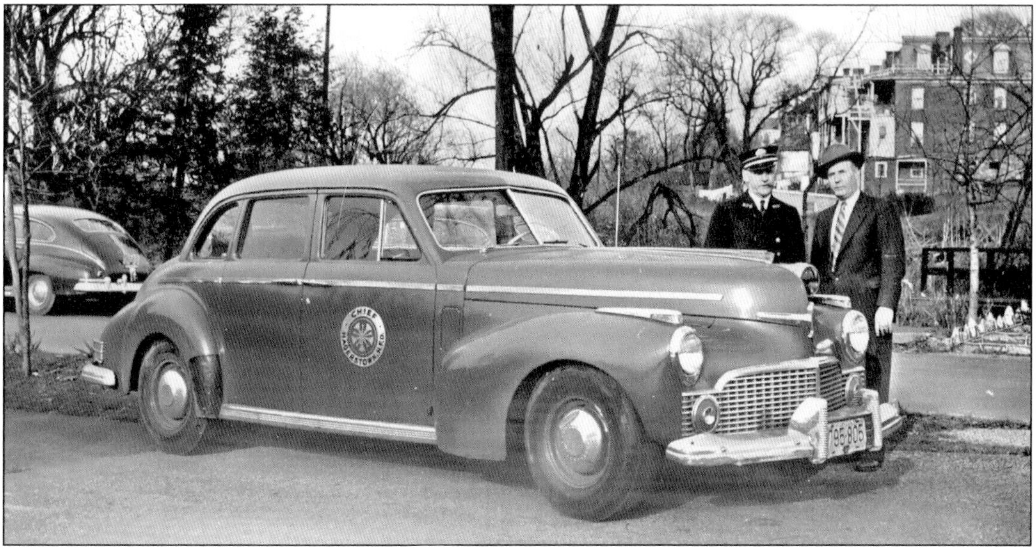

In this photograph is Hagerstown's first paid fire chief, Max Hoover. Chief Hoover (left) was born on May 2, 1891. He was first a volunteer at the Independent Junior Fire Company. From 1929 until 1948, Chief Hoover served as Hagerstown's volunteer chief. On May 1, 1948, he was selected as the paid fire chief. Chief Hoover served until his death on May 1, 1952. Standing next to the chief is First Hagerstown Hose Company member Maurice Domenici. Prior to becoming the chief's car, this Studebaker President was used by Mayor Mills. (Courtesy of the C. William Karn collection.)

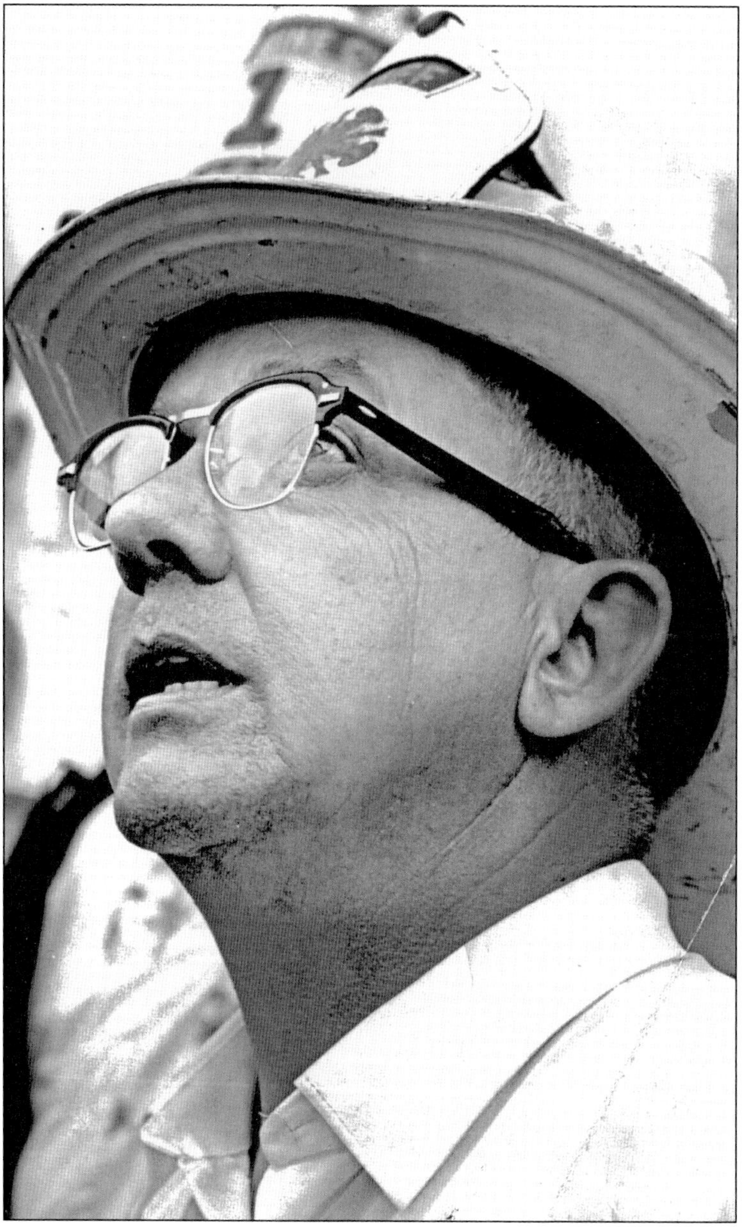

One of the most well-respected men in the history of the Hagerstown Fire Department is fire chief John Hall. Chief Hall earned the respect of his men through dedication, an ability to get things done, and his knack for "telling it the way it is." Chief Hall joined the department as a volunteer at the Western Enterprise Fire Company in 1940. He became a driver at the Pioneer Hook & Ladder Company in 1950, and he served as fire chief from 1952 until 1974. His accomplishments included the establishment of relief drivers and the acquisition of uniforms for the firefighters. One time on a fire incident, a police officer mistook driver William Hovermill for a citizen and made him stand back with the rest of the crowd. Shortly thereafter, uniforms were purchased for all firefighters. This photograph was taken at the Saint Johns Lutheran Church fire at 141 South Potomac Street on July 20, 1969. (Courtesy of the First Hagerstown Hose Company collection.)

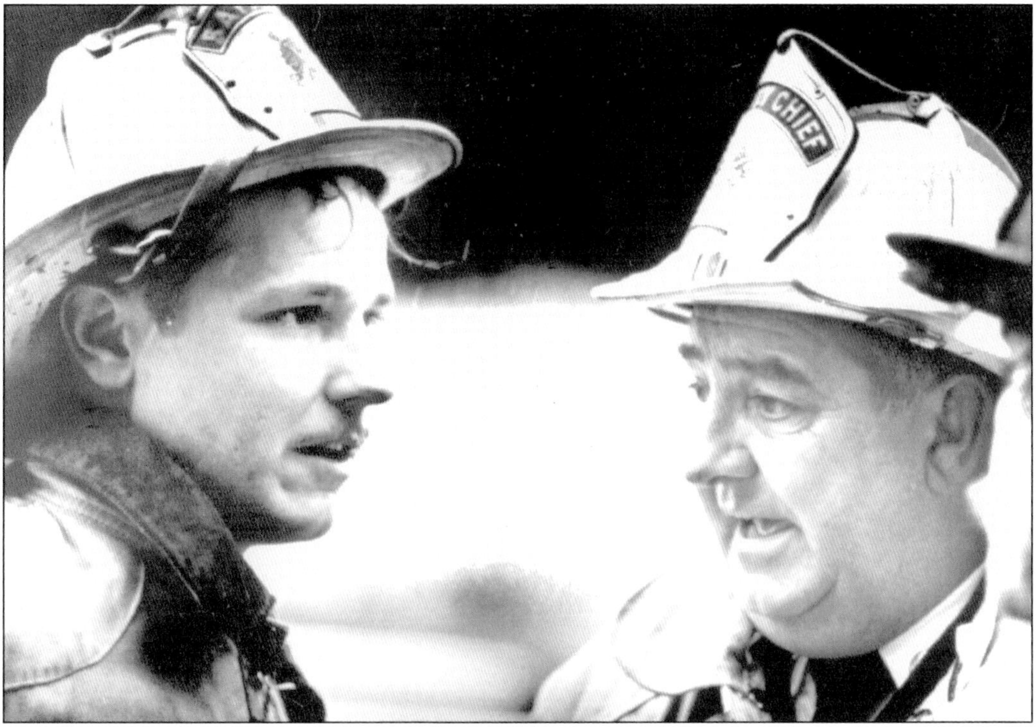

Dep. Fire Chief C. William Karn (right) converses with Batt. Chief C. Kingsley Poole at the scene of a fire. Karn was hired on April 1, 1953. He served as a relief driver and regular driver at the hose company and was promoted to deputy chief in 1969. He served as acting fire chief from 1974 to 1976. Chief Karn retired on February 24, 1984. Chief Poole served as Battalion Chief 2 for 26 years. He retired in 2001 with over 38 years of service to the city as a volunteer firefighter and career officer. (Courtesy of the First Hagerstown Hose Company.)

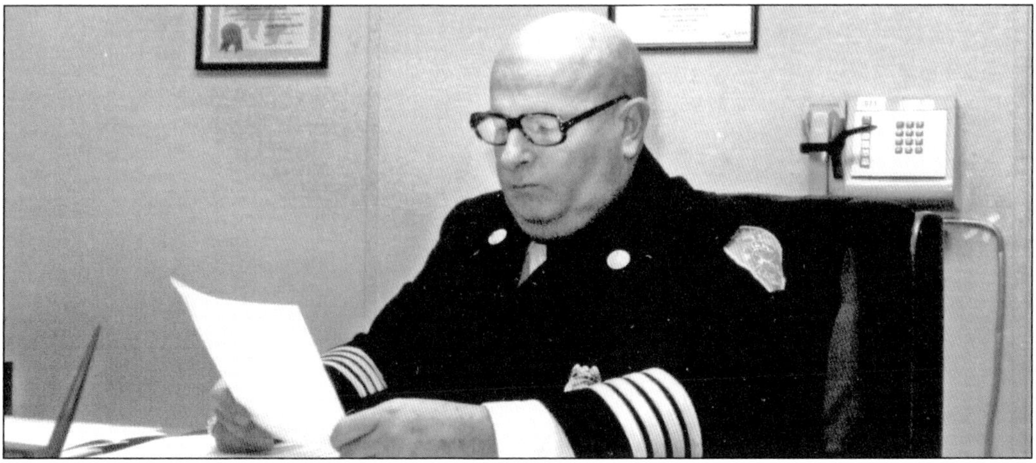

A fire chief's job is never done. Here is Chief Richard Smith reviewing paperwork. A large part of a chief's responsibility consists of handling administrative issues. Chief Smith joined the fire department as a volunteer firefighter at the Pioneer Hook & Ladder Company in 1953. After being hired as a relief driver in 1959, Chief Smith has held every line-officer position, including the rank of fire chief from 1976 until his retirement in 1985. (Courtesy of the Richard Smith collection.)

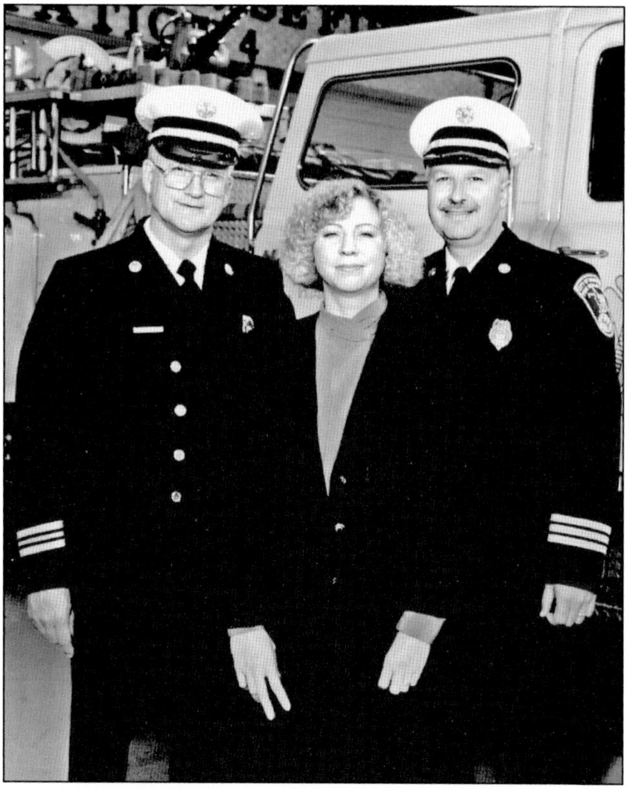

Prior to becoming the Hagerstown fire chief, Gary Hawbaker (right) was the volunteer fire chief at the Rescue Hose Company No. 1, located in Greencastle, Pennsylvania. Chief Hawbaker has held the rank of fire chief from 1985 until the present. Dep. Chief Baker (left) served as a volunteer at South Hagerstown before holding all line-officer positions including acting fire chief in 1985. Administrative secretary Cindy Lewis (center) is the person who makes the office run smoothly. (Courtesy of the Chief Gary Hawbaker collection.)

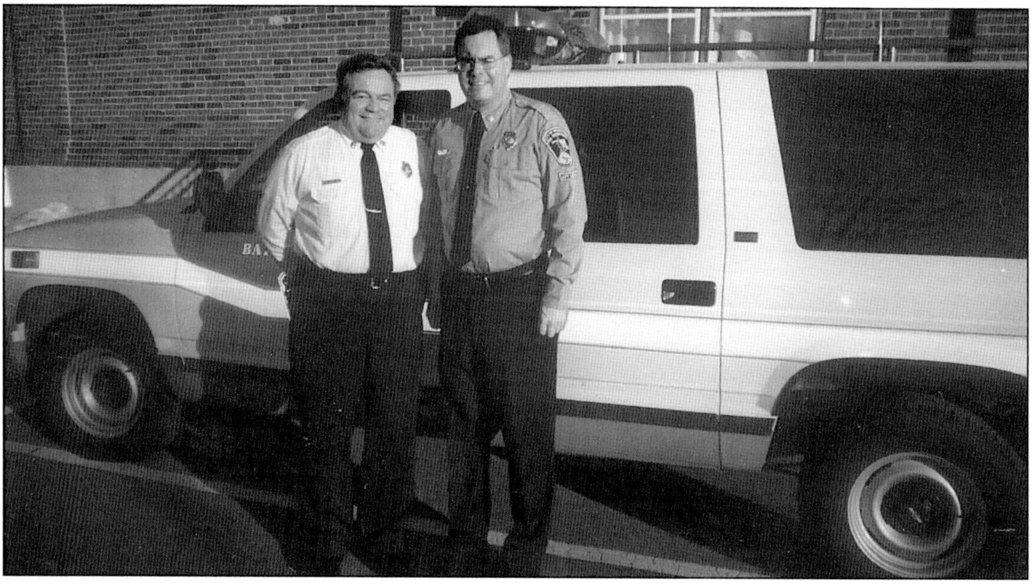

Dep. Chief Richard Kipe and Firefighter/Life Safety Educator Michael Weller pause for this photograph on Chief Kipe's last day at work. Chief Kipe served the city faithfully for over 40 years of combined service as a volunteer firefighter and career officer. Chief Kipe held all line-officer positions including the rank of acting fire chief in 2003. Weller is the city's first life-safety educator. He coordinates all life-safety classes and related activities. Weller joined the department in 1981. (Courtesy of the author.)

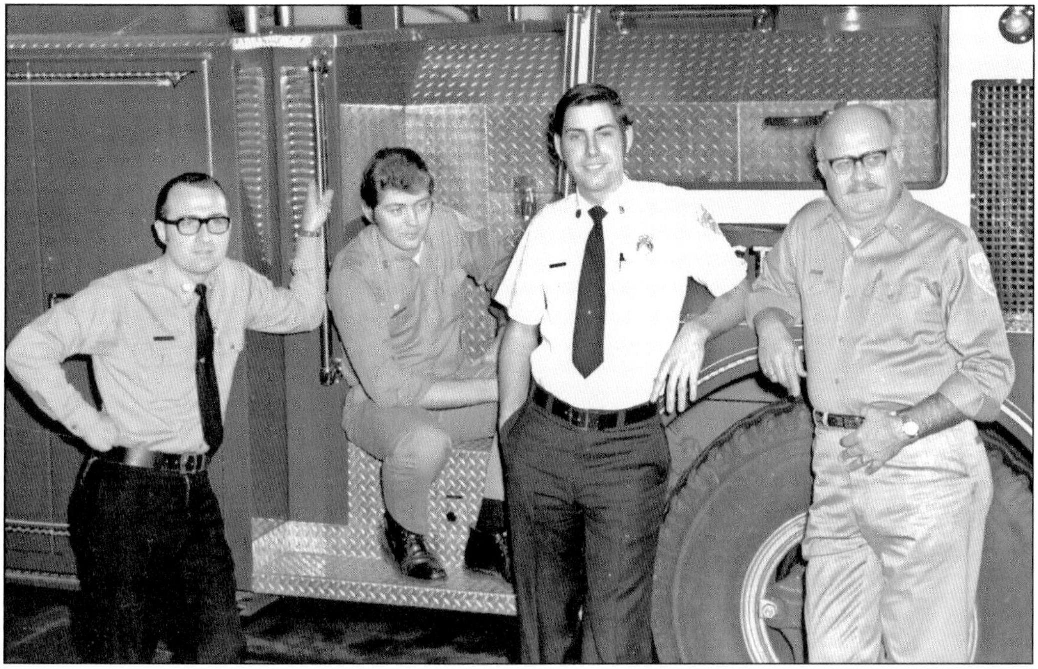

On August 1, 1974, the fire department switched from two shifts working 24 hours at a time to three shifts working two 10-hour days and two 14-hour nights, then two 24-hour days off. Pictured here from left to right are FAO William "Jug" Thompson, FAO Larry Williams, Batt. Chief Steve Brezler, and FAO Charles Garrett. (Courtesy of the Richard Smith collection.)

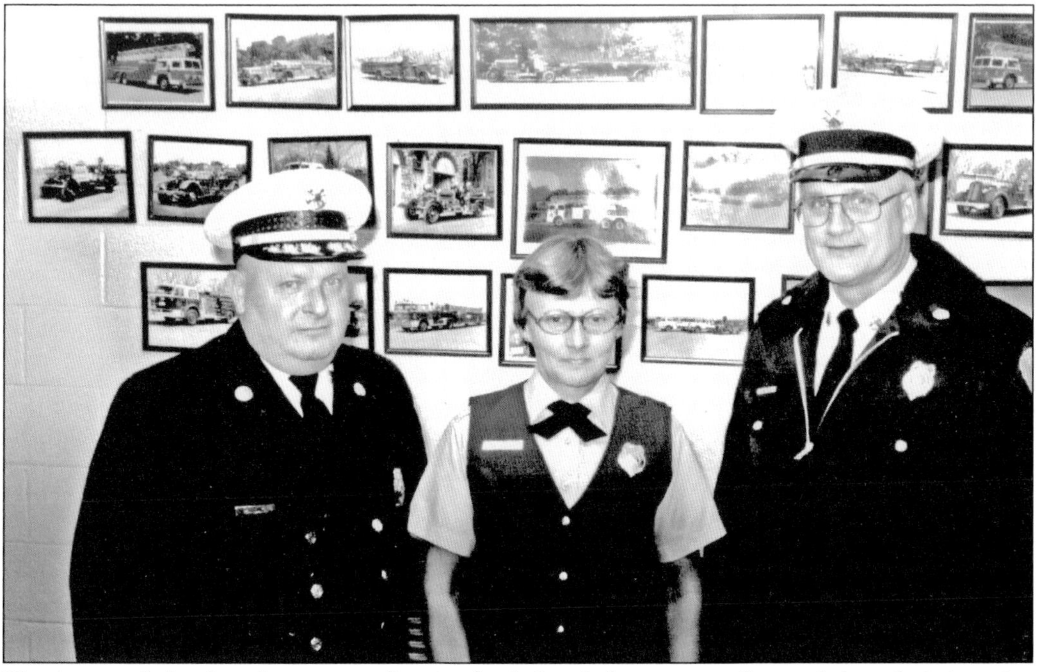

The heart of the fire department lies in the behind-the-scenes efforts of the administrative secretary. Pictured here from left to right are Chief Richard Smith, administrative secretary Carol Cogar, and Dep. Fire Chief Charles Baker. (Courtesy of the Richard Smith collection.)

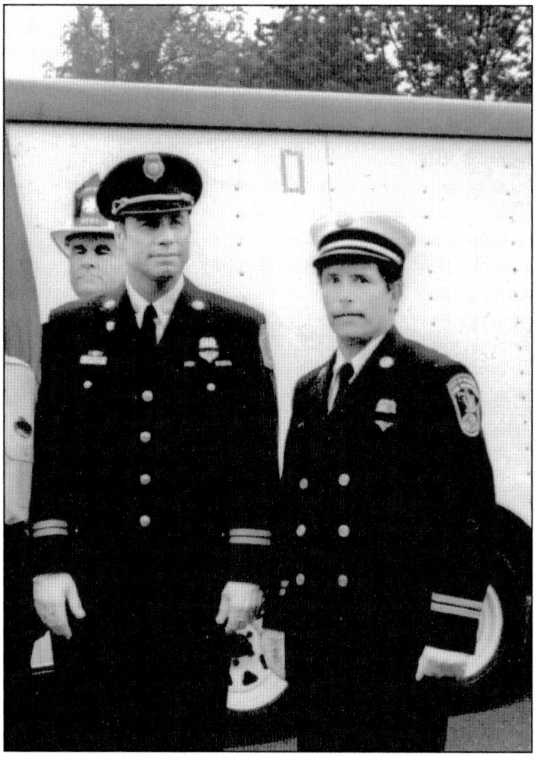

This view shows the scene of the *Ladder 49* movie set. The Hagerstown Fire Department Honor Guard was asked to participate during the filming in Baltimore, Maryland. The author is standing with actor John Travolta. Mr. Travolta plays the role of a Baltimore City Fire Department officer. (Courtesy of Gary Hoffman.)

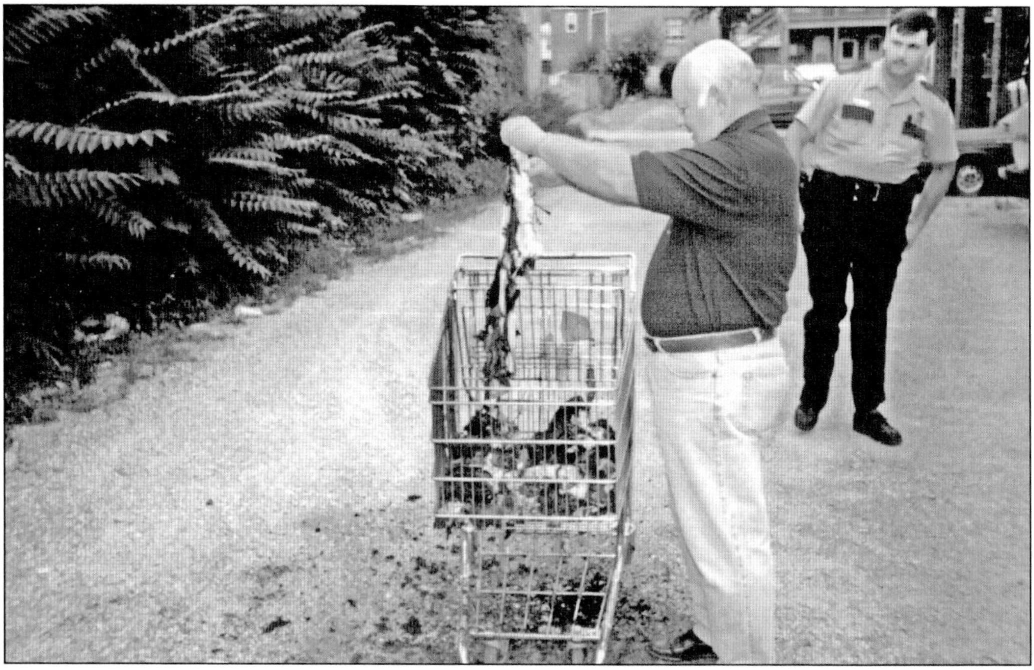

One of the most critical components of the fire department is the fire marshal's office. Asst. Fire Marshal John Hersh is seen here investigating the cause of a small fire on North Locust Street. Hersh is a well-respected investigator and is considered an expert witness in court. Hagerstown police officer Gerry Kendle is standing in the background. (Courtesy of the author.)

Driver Charles Shaffner joined the Independent Junior Fire Company on September 30, 1912. He experienced the privilege of driving the horses and the first motorized apparatus. In this photograph, he is sitting behind the wheel of the Junior's 1915 Seagrave. (Courtesy of the Independent Junior Fire Company.)

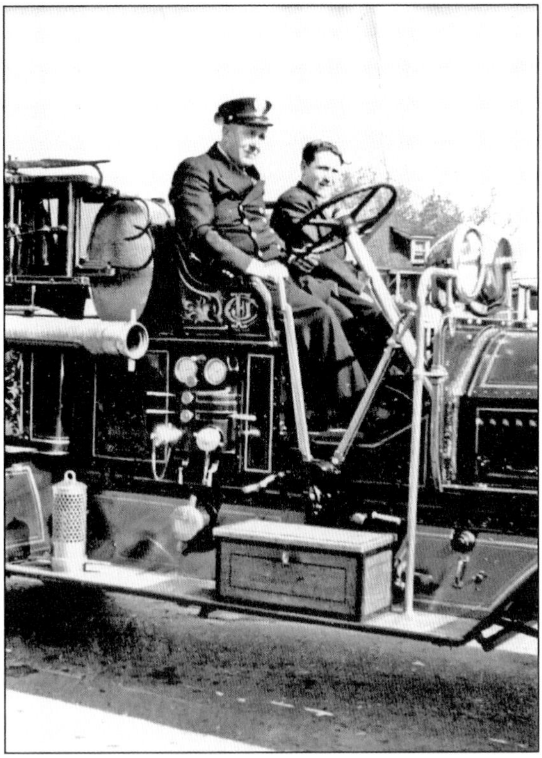

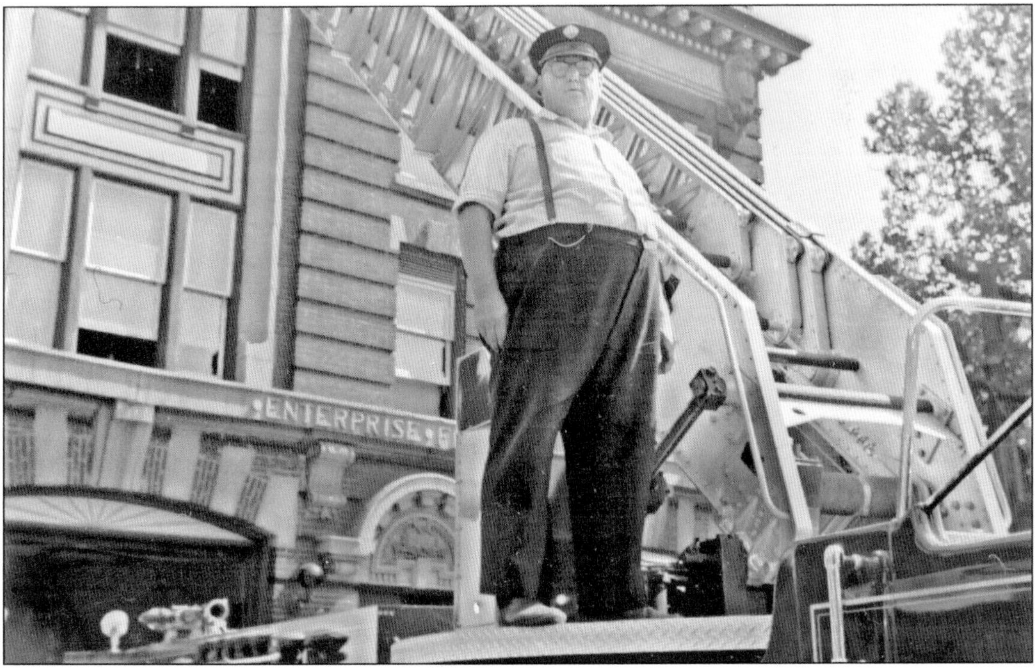

This is a photo of Western Enterprise driver Russell "Peep Sight" Anderson. Anderson began his career as a firefighter in 1918. Born in 1893, he lived in Hagerstown all of his life. Here he is standing on the turntable of the new 1956 Pirsch ladder truck (Truck 4). (Courtesy of the C. William Karn collection.)

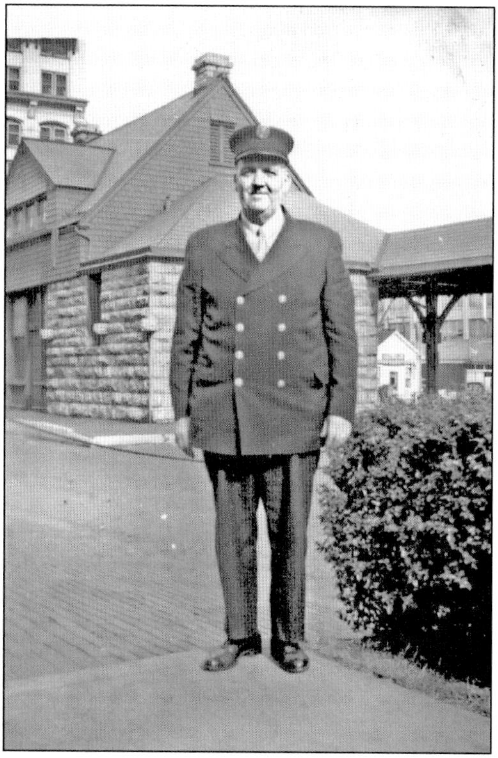

Antietam Fire Company driver John Bomberger was born on December 1, 1888. His fire service career began in 1906. His previous occupation was that of a pipe maker at the Moller Organ works. The B&O Railroad passenger station, located across from the Antietam Fire Company, is in the background. (Courtesy of the Antietam Fire Company.)

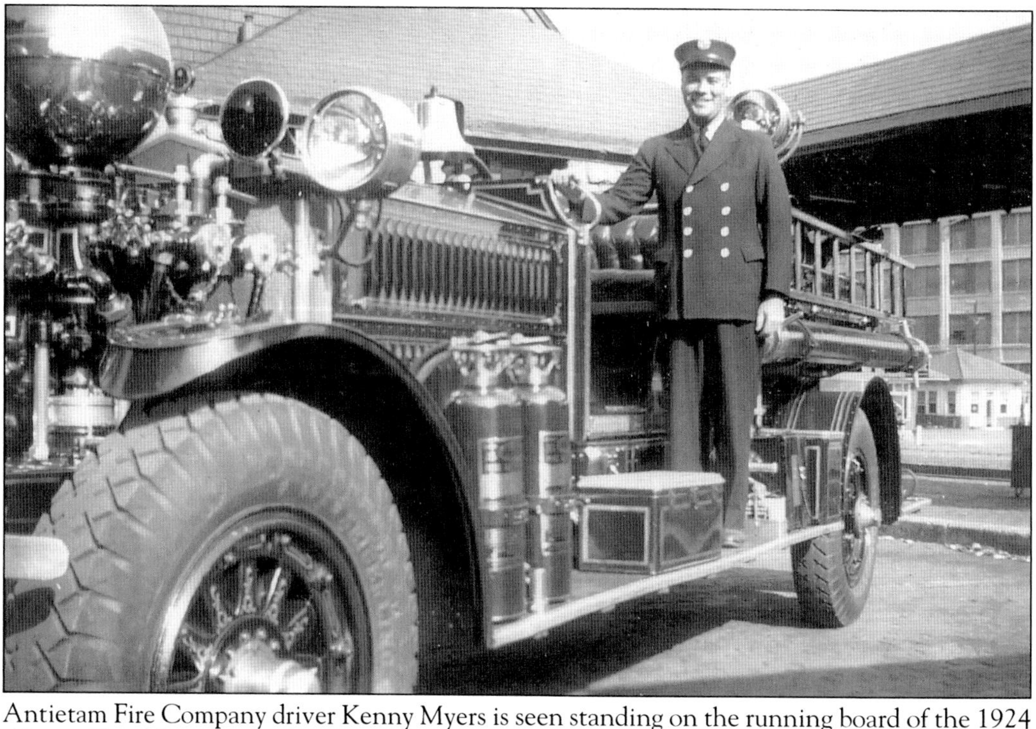

Antietam Fire Company driver Kenny Myers is seen standing on the running board of the 1924 Ahrens-Fox (Queen Mary). The photograph was taken in 1942. (Courtesy of the Antietam Fire Company.)

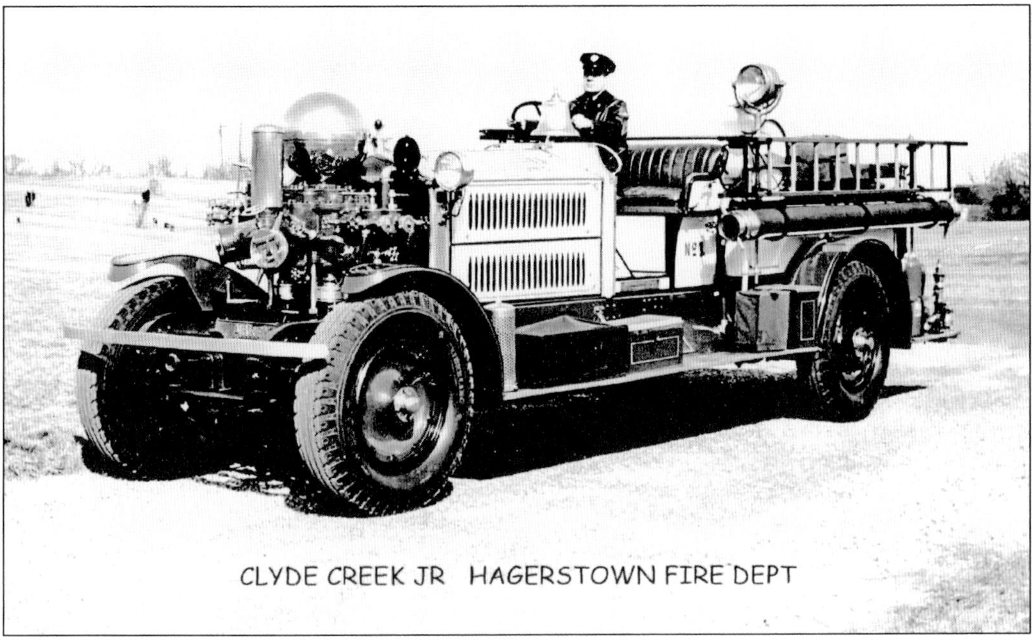

Driver Clyde Creek is shown sitting behind the steering wheel of the First Hagerstown Hose Company's 1923 Ahrens-Fox pumper, "Dynamite." Note the steering wheel is on the right side and Dynamite is painted white. (Courtesy of the William Dieterich collection.)

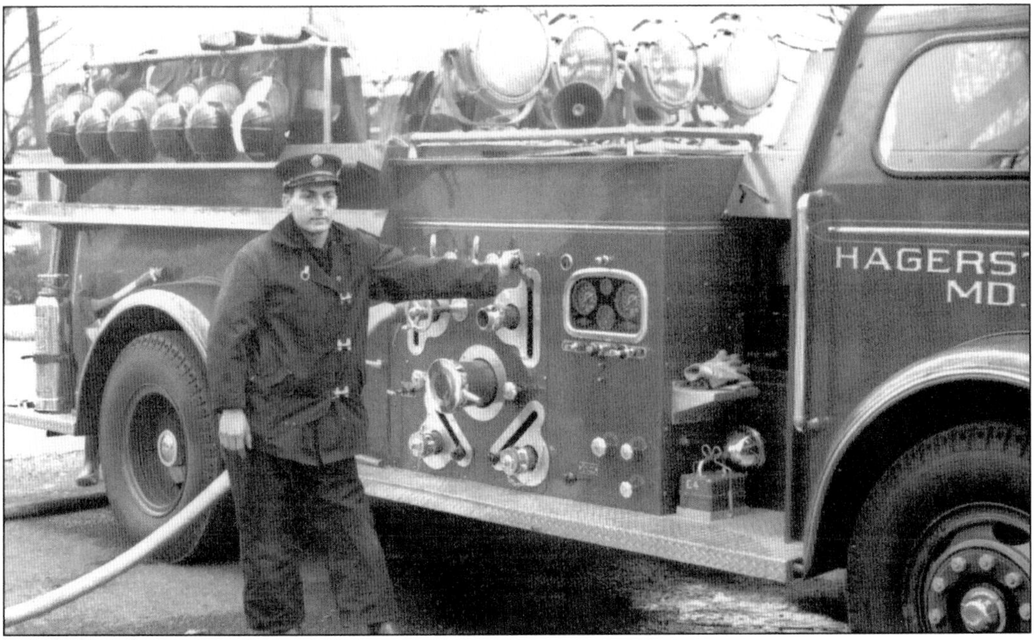

In this February 1959 photograph stands driver William I. Dieterich, who served as a driver from 1950 to 1959 before transferring to the Hagerstown Signal Department. Within the signal department he was responsible for maintaining the Gamewell fire alarm system. Dieterich has served as a volunteer with the Antietam Fire Company, as captain of the Pioneer Hook & Ladder Company, and as chief of the South Hagerstown Fire Company. (Courtesy of the William Dieterich collection.)

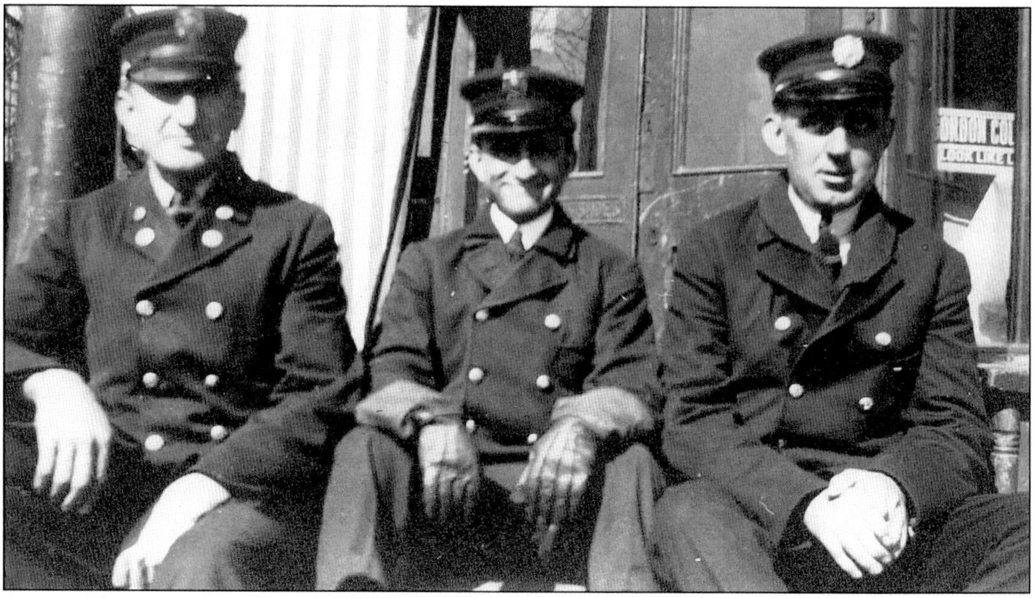

Pictured here are three drivers of the Pioneer Hook & Ladder Company. From left to right are "Polly" Lohr, George E. "Pauly" Paulsgrove, and Charles J. Flook Sr. Paulsgrove was formerly a cigar maker. He entered public service as a firefighter in 1914. (Courtesy of the First Hagerstown Hose Company.)

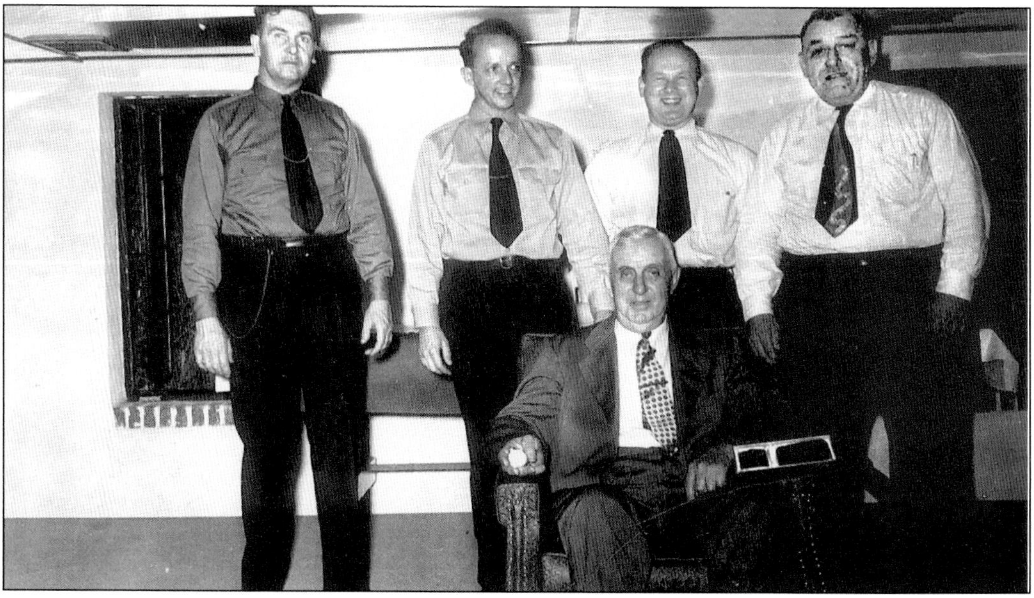

Attending the retirement party of Pioneer Hook & Ladder driver Charles Flook Sr. (seated) are, from left to right, drivers William Beard, John Hall (future chief), Elwood Higgs, and Herman Negley. Flook worked for the fire department from 1914 until 1917. He returned to the fire department in 1920 after operating a garage for 10 months and working for the railroad. (Courtesy of the Richard Smith collection.)

Representing the IAFF Local 1605 scholarship committee are FAO Ben Conrad (left) and Lt. Richard Smith (union president at the time) (center). They are shown presenting Carl J. Galligan of Hagerstown Junior College with a scholarship check on January 21, 1971. The check represents proceeds from the first annual vaudeville show. (Copyright *The Herald-Mail* Company, Hagerstown, Maryland.)

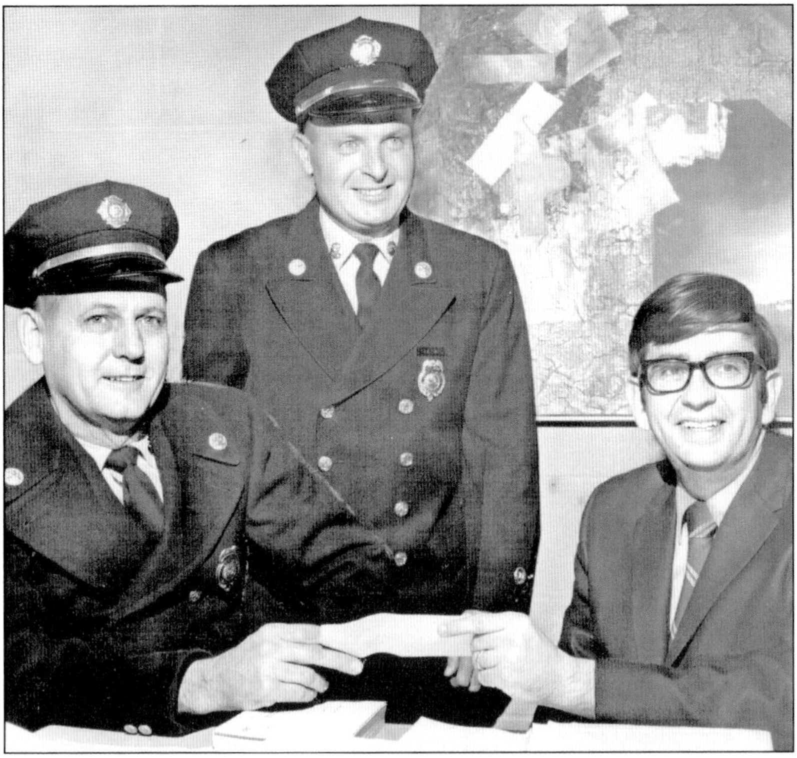

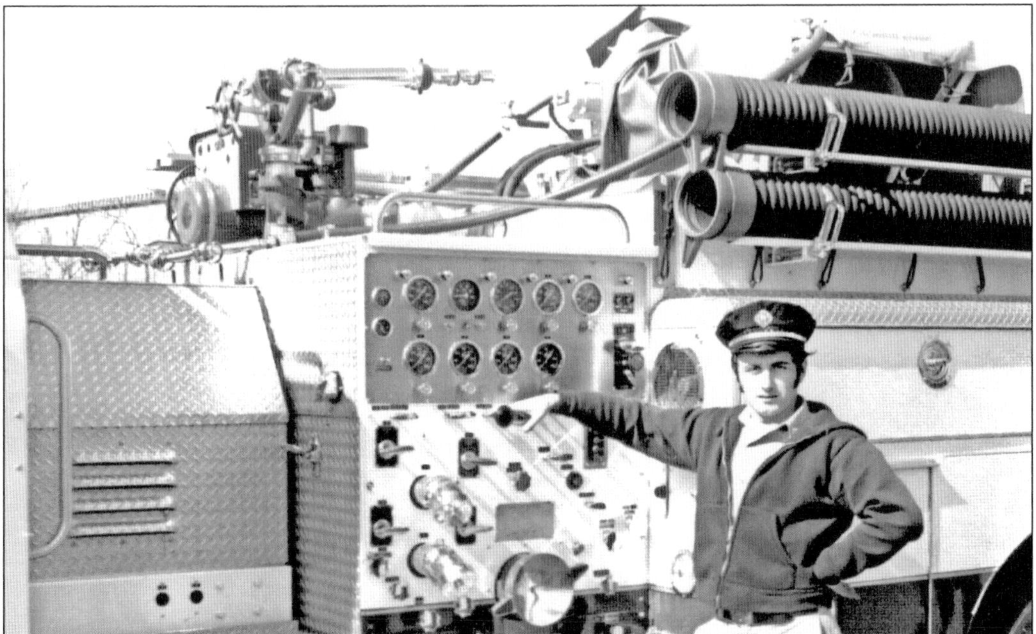

FAO Richard Jordan operates the pump of South Hagerstown's 1972 Mack pumper. The photograph was taken on February 22, 1975, at the Hagerstown Training Center on Bowman Avenue. FAO Jordan became a career firefighter on April 22, 1974. (Courtesy of the Kyd Dieterich collection.)

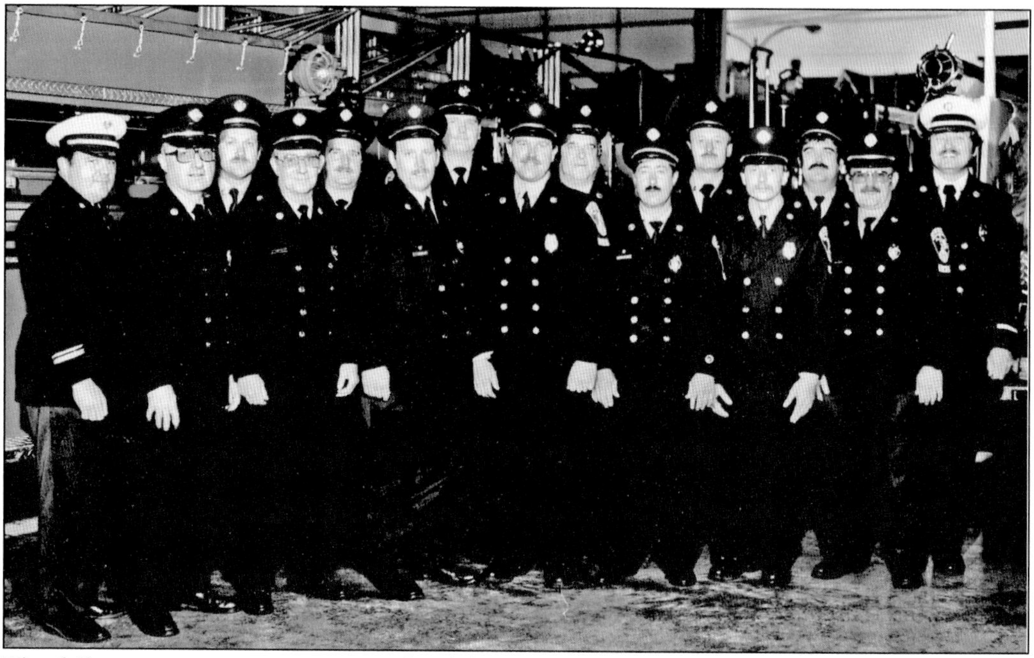

The following pictures represent each of the three shifts that made up the Hagerstown Fire Department in 1991. In the "A" shift, from left to right, are (front row) Batt. Chief Richard Kipe, William Thompson, Herbert Sprecher, Joseph Reichert, C. Edward Shindle, Jerry Cunningham, Tim Shafer, and Larry Alsip; (back row) Leonard Shockey, James Stevens, Richard Conrad, Charles McCoy, Ronald Randall, Robert Herr, and Capt. Ronald Horn. (Courtesy of the Edward Shindle collection.)

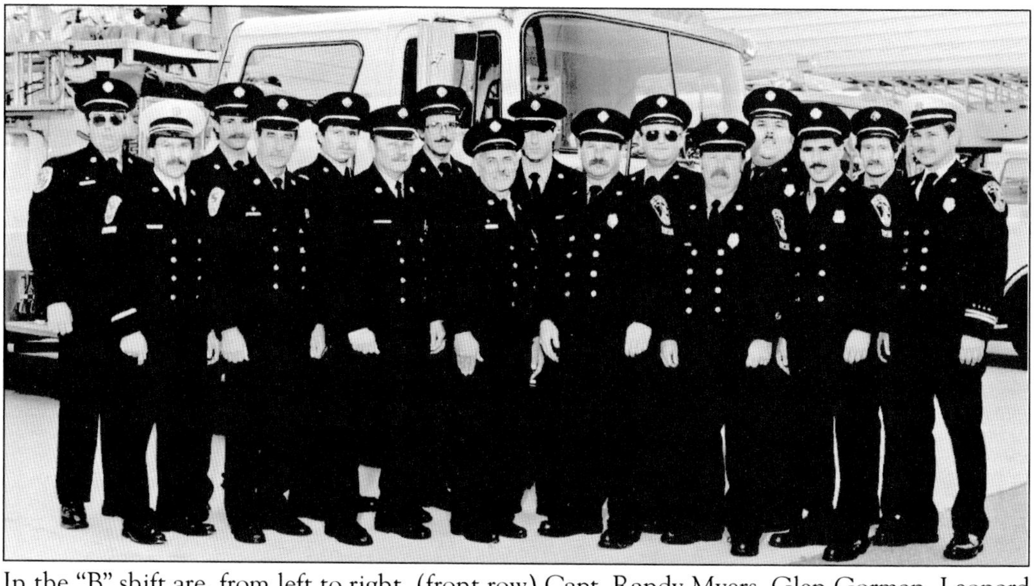

In the "B" shift are, from left to right, (front row) Capt. Randy Myers, Glen Gorman, Leonard Barton, Blaine Carbaugh, David Baer, Gary Hoffman, Richard Gilbert, and Batt. Chief C. Kingsley Poole; (back row) Michael Ward, Vinson Dieterich, Richard Miller, Larry Koons, Douglas DeHaven, Frank Ridenour, Alan Gladhill, and John Thomas. (Courtesy of the C. William Karn collection.)

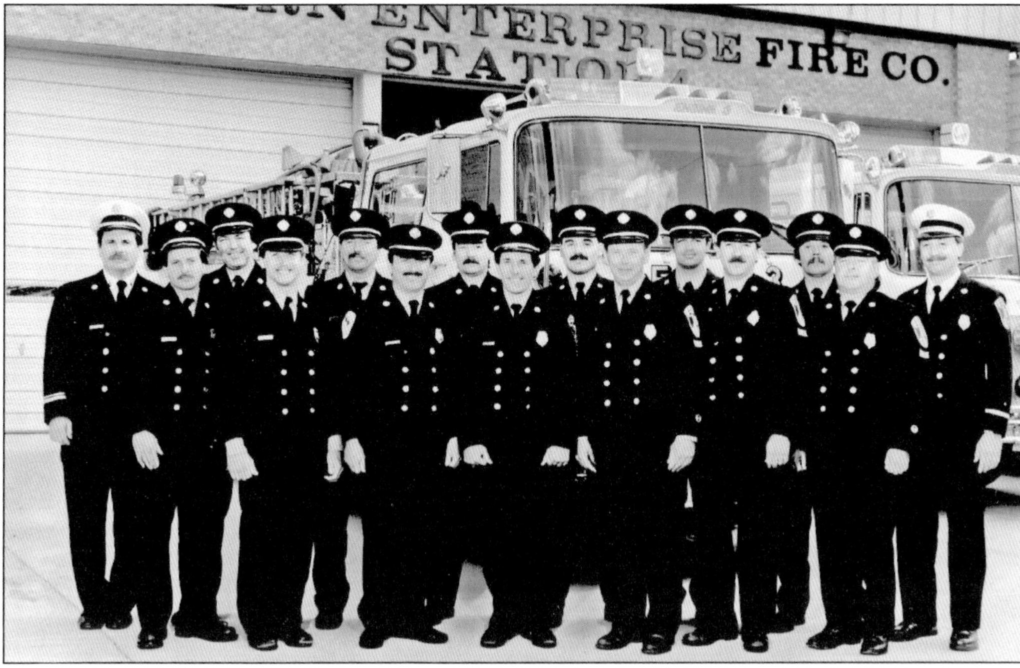

In the "C" shift are, from left to right, (front row) Randy Hinkle, David Poffenberger, Richard Jordan, Justin Mayhue, Mitchell Gearhart, Michael Shade, and Robert Manley; (back row) Batt. Chief Richard Hopkins, Larry Williams, J. Andrew Hartman, Brian Pile, Russell Grim, Kevin Eichelberger, Warren Keyfauver, and Capt. W. Kyd Dieterich. (Courtesy of the author.)

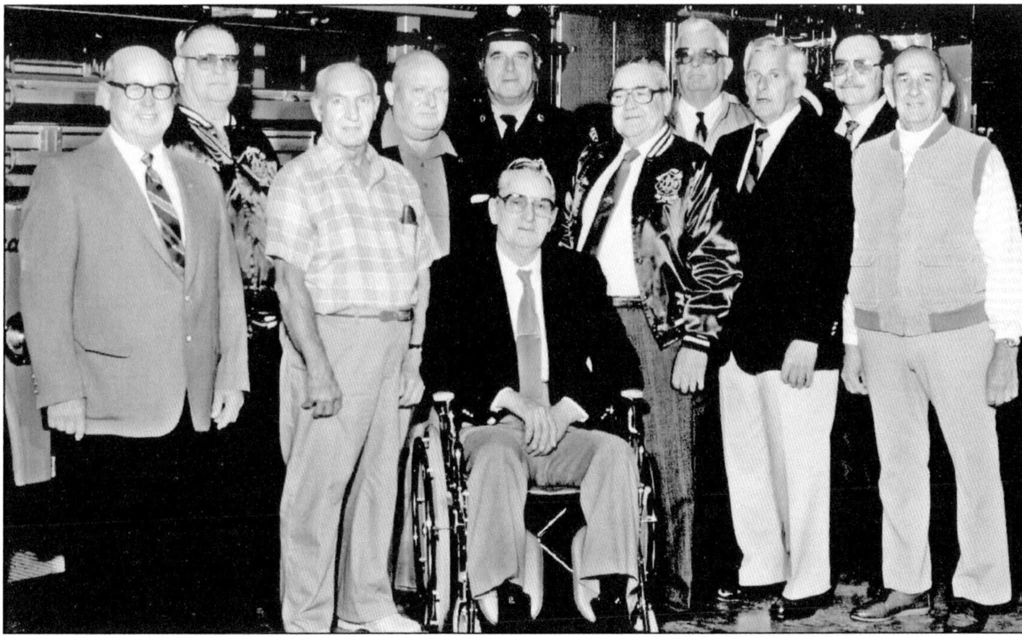

The IAFF Local 1605 was organized in 1966. Shown here are some of the charter members of the organization. From left to right are Elwood Higgs, Harry Daveler, Richard Trovinger, Richard Smith, Edward Elgin (seated), Charles McCoy, Edgar Delinger, Charles Flook, William Kershner, Ronald Moser, and Henry DeLauney. (Courtesy of the Harry Daveler collection.)

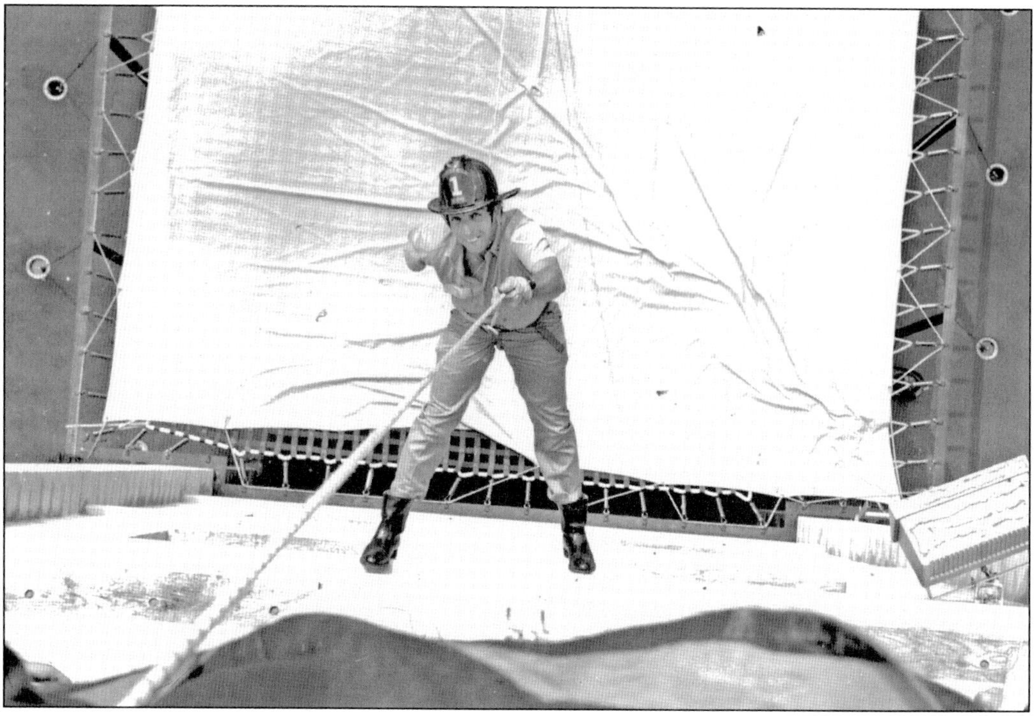

Here is Richard "Shakey" Cramer at the Hagerstown Training Center tower in 1974. FAO Cramer began his career on February 16, 1966. He retired after 17 years of dedicated service in May 1983. (Courtesy of the Richard Smith collection.)

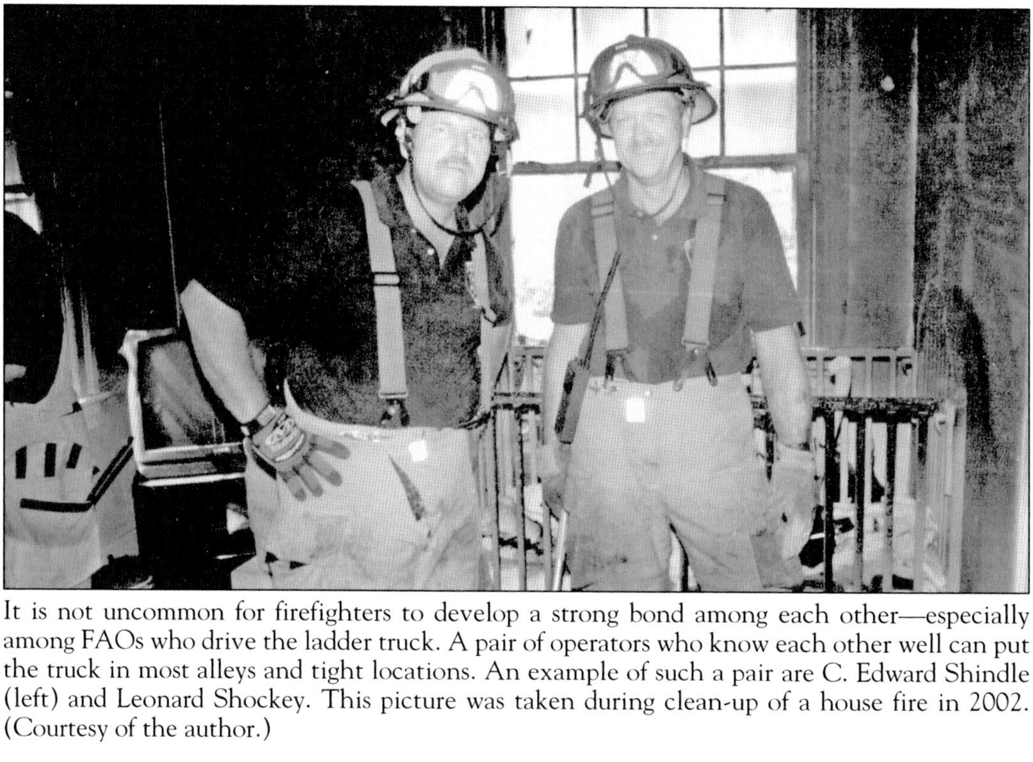

It is not uncommon for firefighters to develop a strong bond among each other—especially among FAOs who drive the ladder truck. A pair of operators who know each other well can put the truck in most alleys and tight locations. An example of such a pair are C. Edward Shindle (left) and Leonard Shockey. This picture was taken during clean-up of a house fire in 2002. (Courtesy of the author.)

Eight

Sound the Alarm

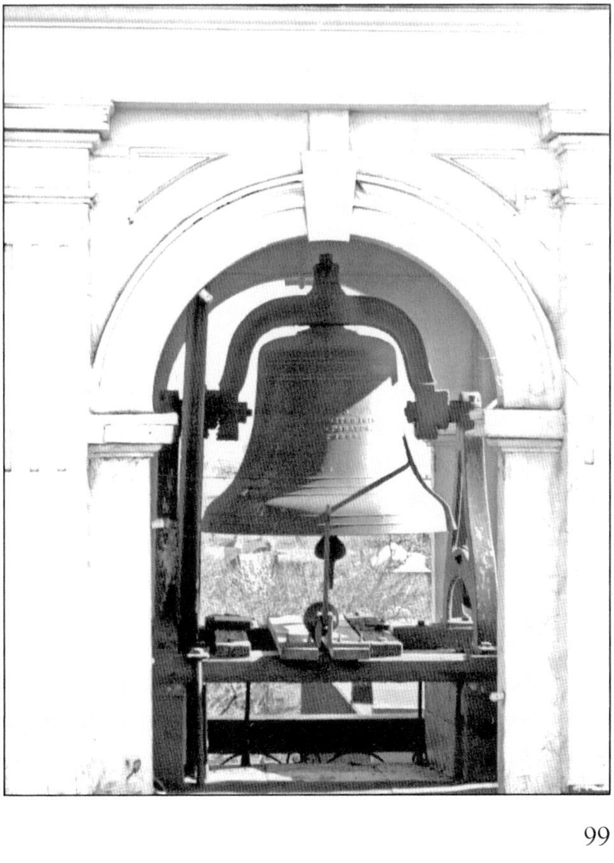

This bell was purchased on May 2, 1884, from the McShane Bell Company of Baltimore, Maryland. It still towers over the First Hagerstown Hose Company. The bell weighs 1,500 pounds. It was named "Rufus" after Rufus M. Hays, the chief projector of the project. The cost of the bell was $340. (Courtesy of the First Hagerstown Hose Company.)

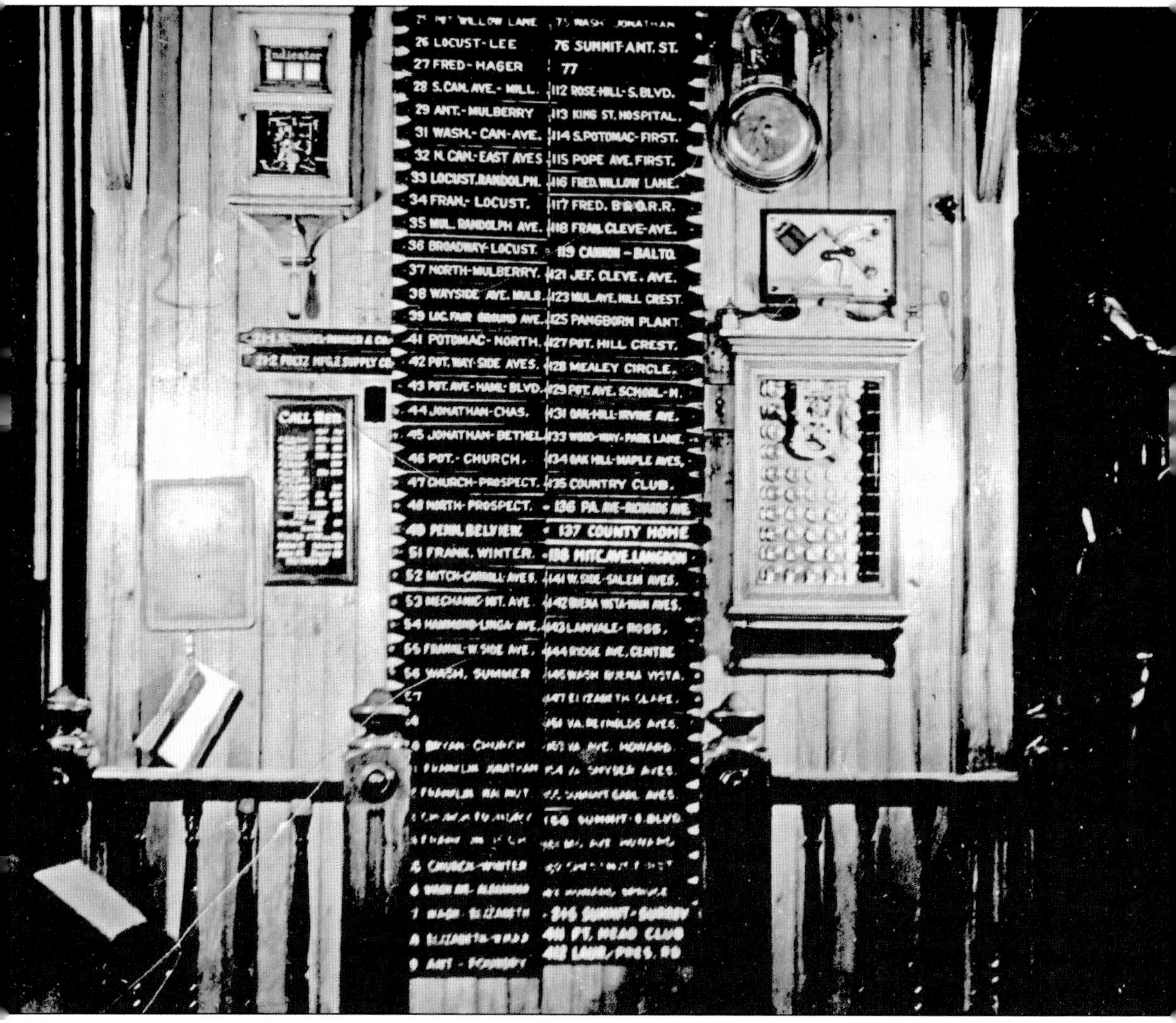

The Gamewell fire alarm system went into service in Hagerstown in 1895. The main alarm board was located at the First Hagerstown Hose Company. Alarms could be received or transmitted from this location. The box on the right is the Gamewell transmitter. After receiving a request for the fire department, the driver/dispatcher would place the appropriate box-alarm code wheel on the transmitter. A lever was pulled that activated the system. The alarm would then be received by each city firehouse. The box on the upper left was the indicator or receiver. If a citizen activated a street box, the alarm would be received by each firehouse simultaneously. This type of transmitter and indicator was replaced by a more modern transmitter and punch-type register in about 1939. (Courtesy of the First Hagerstown Hose Company.)

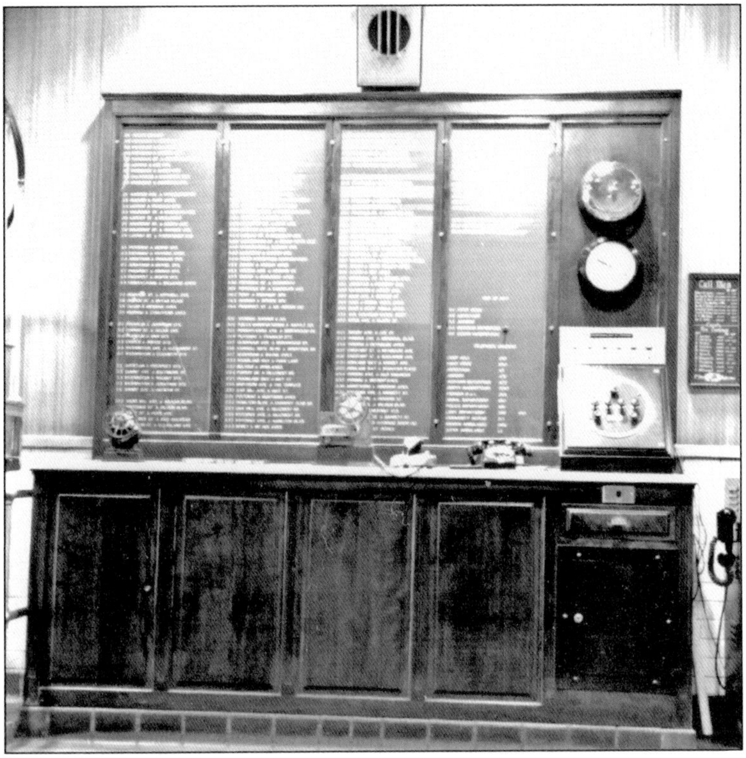

This is the alarm board at the hose company after renovations. The code-wheel transmitter was replaced by a digital transmitter that could transmit any numbers between 1 and 9999. The tape register and take-up reel (center) replaced the indicator box. (Courtesy of the First Hagerstown Hose Company.)

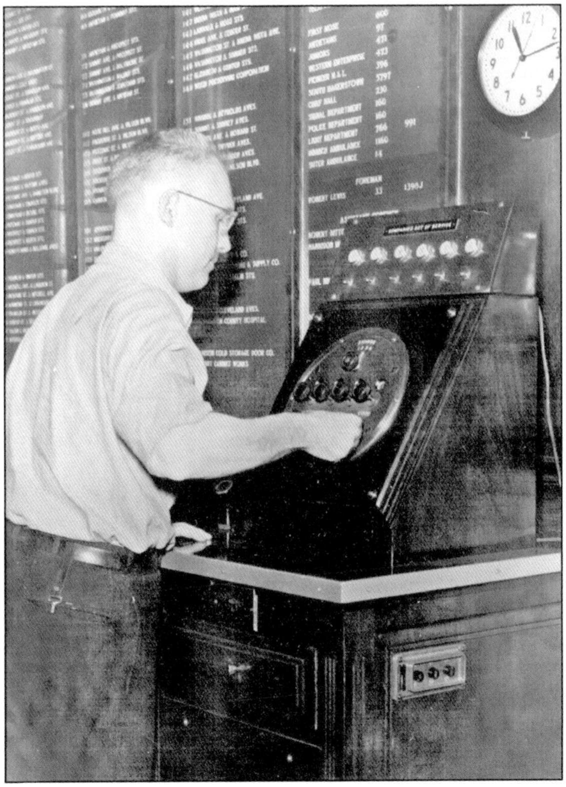

When a citizen called in a fire, the driver/dispatcher would set up the box number closest to the address of the fire. He would then notify the city firehouses via an intercom system (vocal alarm). The Gamewell transmitter was then activated. If the hose company was due on the call, volunteer firefighters would often stand by to handle additional incidents. This is Driver/Dispatcher A.K. McGraw transmitting Box 4–5 located in Harmons Alley on February 17, 1954. (Courtesy of the First Hagerstown Hose Company.)

FIRE ALARMS

Box	Location
2-1	Public Square
2-2	Pot. & Antietam
2-3	Pot. & Baltimore
2-4	Locust Point & Pot.
2-5	Pot. & Willow Lane
2-6	Locust & Lee
2-7	Frederick & Hager
2-8	S. Can. Av. & Mill
2-9	Ant. & Mulberry
3-1	Wash. & Can. Ave.
3-2	N. Can. & East Ave.
3-3	Locust & Randolph
3-4	Franklin & Locust
3-5	Mul. & Rand. Ave.
3-6	Broadway & Locust
3-7	North & Mulberry
3-8	Wayside & Mul..
3-9	Locust & Fair. Ave.
4-1	Potomac & North
4-2	Pot. & Way. Ave.
4-3	Pot. & Ham. Blvd.
*4-4	Jonathan & Chas.
*4-5	Jonathan & Bethel
4-6	Potomac & Church
*4-7	Church & Prospect
*4-8	North & Prospect
*4-9	Penn. & Bel. Ave.
*5-1	Franklin & Winter
*5-2	Mit. & Car. Aves.
*5-3	Mech. & Mit. Ave.
*5-4	Ham. & Ling. Ave.
*5-5	Frank. & West Side
*5-6	Wash. & Summer
*5-9	Bryan & Church
6-1	Franklin & Jonathan
*6-2	Frank. & Walnut
*6-3	Church & Foundry
*6-4	Franklin & High
*6-5	Church & Winter
*6-6	Wash. & Alexander
*6-7	Wash. & Elizabeth
*6-8	Eliz. & W. M. Ry.
*6-9	Ant. & Foundry
7-1	Antietam & Prospect
7-2	Summit & Prospect.
7-3	Sum. Ave. & Balt.
*7-4	Wash. & Walnut
7-5	Wash. & Jonathan
7-6	Ant. & Summit Ave.
1-1-2	Rose H. & S. Blv.
1-1-3	Hospital & King St.
1-1-4	S. Potomac & First
1-1-5	Pope & First St.
1-1-6	Fred. & Willow La.
1-1-7	Fred. & B. & O.
1-1-8	Franklin & Clev.
1-1-9	Cannon & Balt. St.
1-2-1	Jef. & Clev. Ave.
1-2-3	Mul. & Hill Crest
1-2-5	Pangborn Plant
1-2-7	Pot. & Hill Crest
1-2-8	Mealey Circle
1-2-9	Pot. Ave. School
1-3-1	Oak Hill & Irvin
*1-3-3	W. Way & Park L.
1-3-4	Oak Hill & Maple
1-3-5	Country Club
*1-3-7	Mit. & Langdon
*1-4-1	W. Side & Salem
*1-4-2	B. Vista & Main
*1-4-3	Lanvale & Ross
*1-4-4	Ridge Av. & Cen.
*1-4-5	Wash. & B. Vista
*1-4-7	Elizabeth & Clair
1-5-1	Va. & Reynolds
1-5-3	Virginia & Howard
1-5-4	Virginia & Snyder
1-5-5	Sum. & Gar. Aves.
1-5-6	Summit & S. Blvd.
1-6-1	Md. & Howard
1-6-2	Chestnut & First
1-6-3	Howard & Spruce
2-1-1	Schindel R. & Co.
2-1-2	Foltz Mfg Co.

*Calls not answered by Hose Co.

This is a pocket-size version of a fire-alarm box locator. Before 1939, when the Gamewell system was activated either by a citizen pulling a street box or the driver/dispatcher at the hose company, a bell called "Big Six" (located atop City Hall and also atop the Western Enterprise Fire Company) would ring. City volunteer firefighters would count the number of taps, locate the alarm on the pocket card, and respond to the address. In 1939 the Enterprise bell was replaced by a diaphone air horn. In 1951 an air horn was installed at the South Hagerstown Fire Company. The Big Six bell was replaced by an air horn *c.* 1969. (Courtesy of the First Hagerstown Hose Company.)

On July 13, 1959, the Washington County Hospital Box 291 was changed to Box 651. In Hagerstown, any box beginning with a six is considered a master box. A master box is capable of receiving internal alarms such as pull stations and sprinkler alarms and then sending the alarm to the firehouses. Changing the box here are signal department employees William Dieterich (left) and Dick Mentzer. (Courtesy of the First Hagerstown Hose Company.)

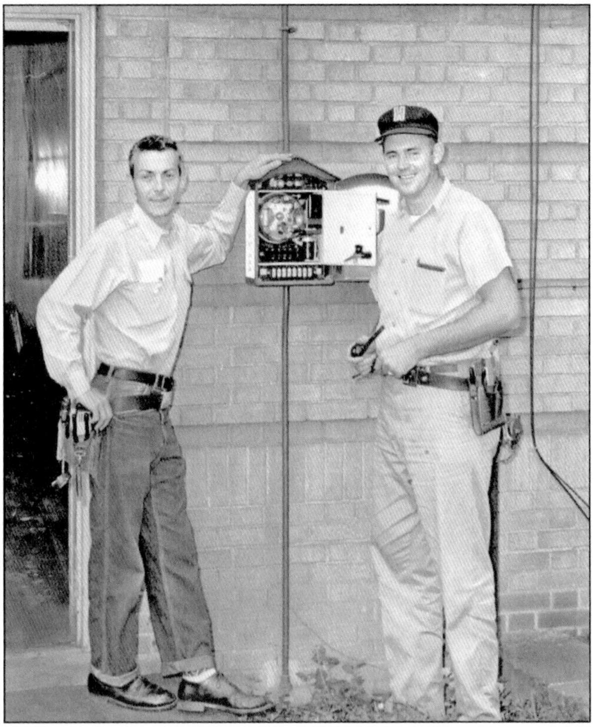

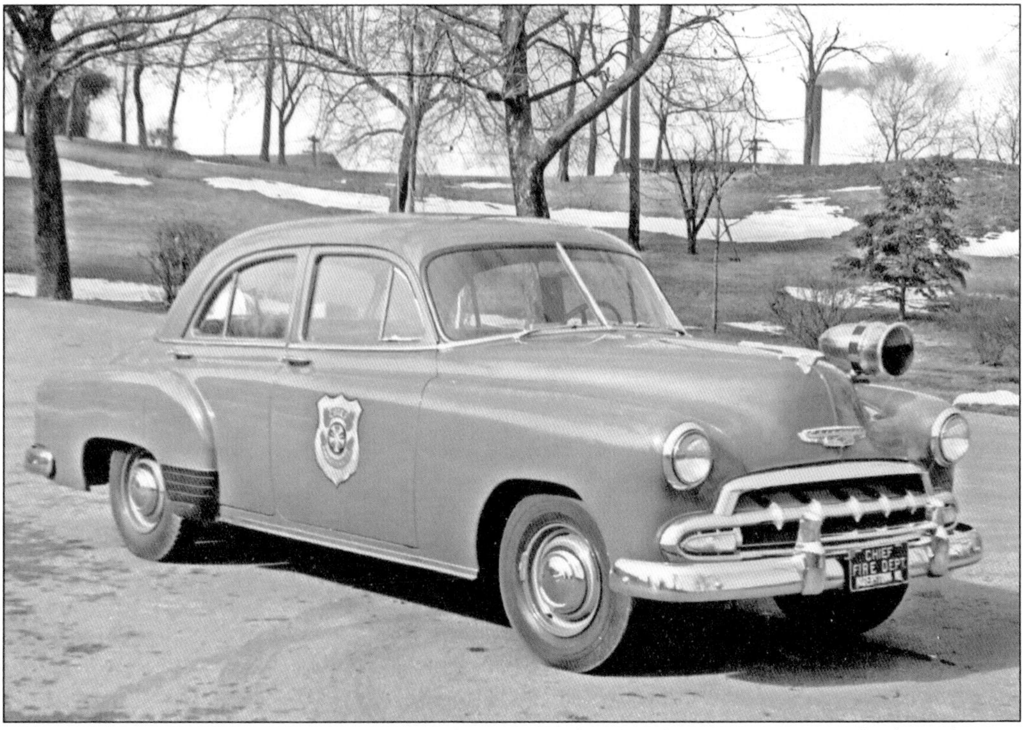

This was Chief Hall's car, a 1951 Chevrolet. It was the first fire department vehicle to have a two-way radio. The chief could only communicate with the police department until additional radios were installed on the fire apparatus. (Courtesy of the C. William Karn collection.)

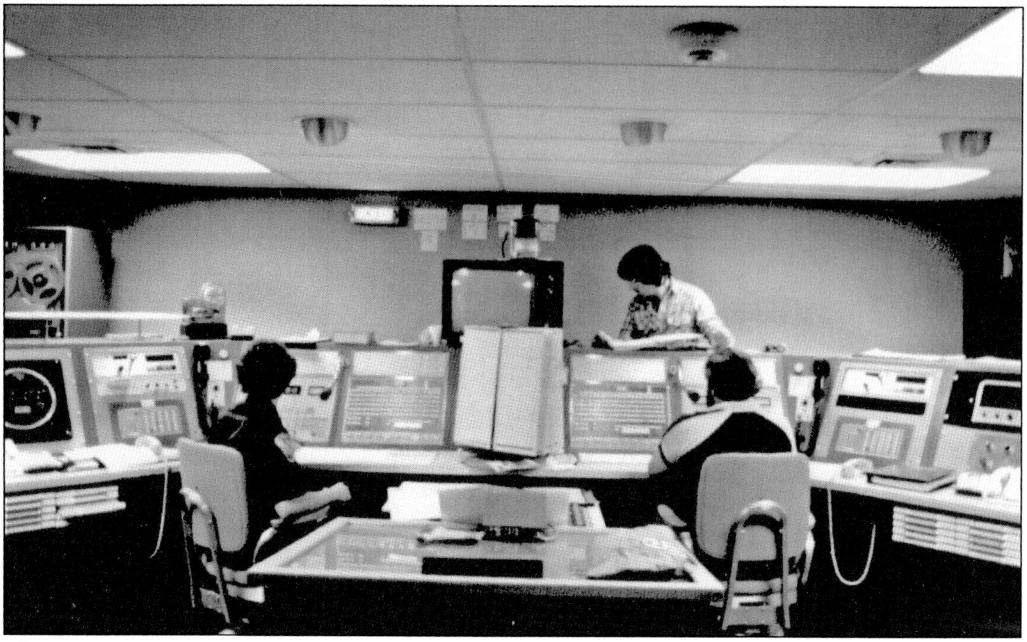

This is an overview of the original console of the Washington County Fire and Rescue Communications Center (central alarm). This center, located at 33 West Washington Street, replaced the alarm center at the hose company. The dispatcher on the left is Robert S. Kefauver, who dispatched the first call on August 24, 1975, at 137 John Street for two injured persons. The dispatcher on the right is Darren Andrews. The communications maintenance technician, Pete Lowenheim, is standing behind the console. (Courtesy of Fire and Rescue Communications.)

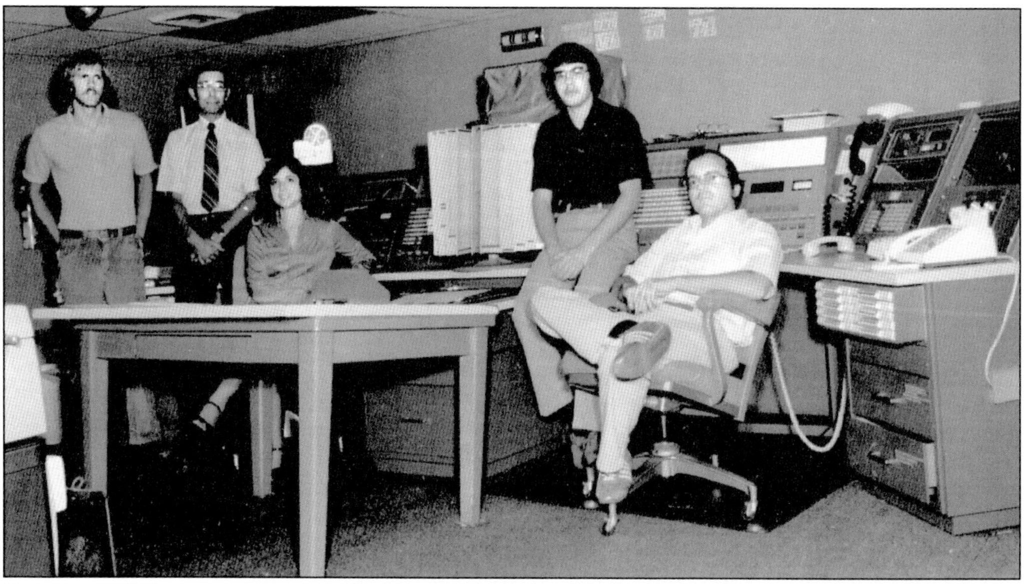

Represented here are some of the employees responsible for the early success of the Communications Center. From left to right are Michael Harpster (communications maintenance technician), Chief Ronald Karn, Verna Brown (administrative secretary), Michael Weller (dispatcher), and Dep. Chief Sidney Mills. (Courtesy of the First Hagerstown Hose Company.)

Nine

A Firefighter's Friend

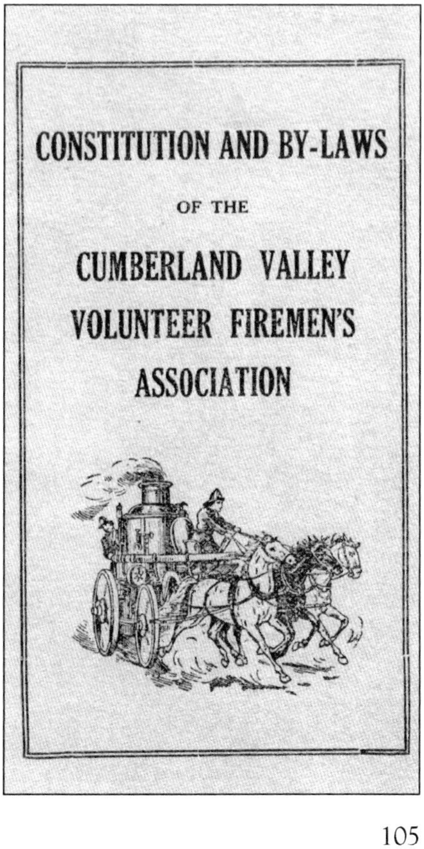

In 1901, J.F. Boyd, the superintendent of the Cumberland Valley Railroad, invited various fire department officers to Chambersburg, Pennsylvania, to advise and adopt a uniform fire-hose coupling thread. The first meeting brought such a favorable impression that a permanent organization was formed. A convention is held annually in the region. (Courtesy of the Independent Junior Fire Company.)

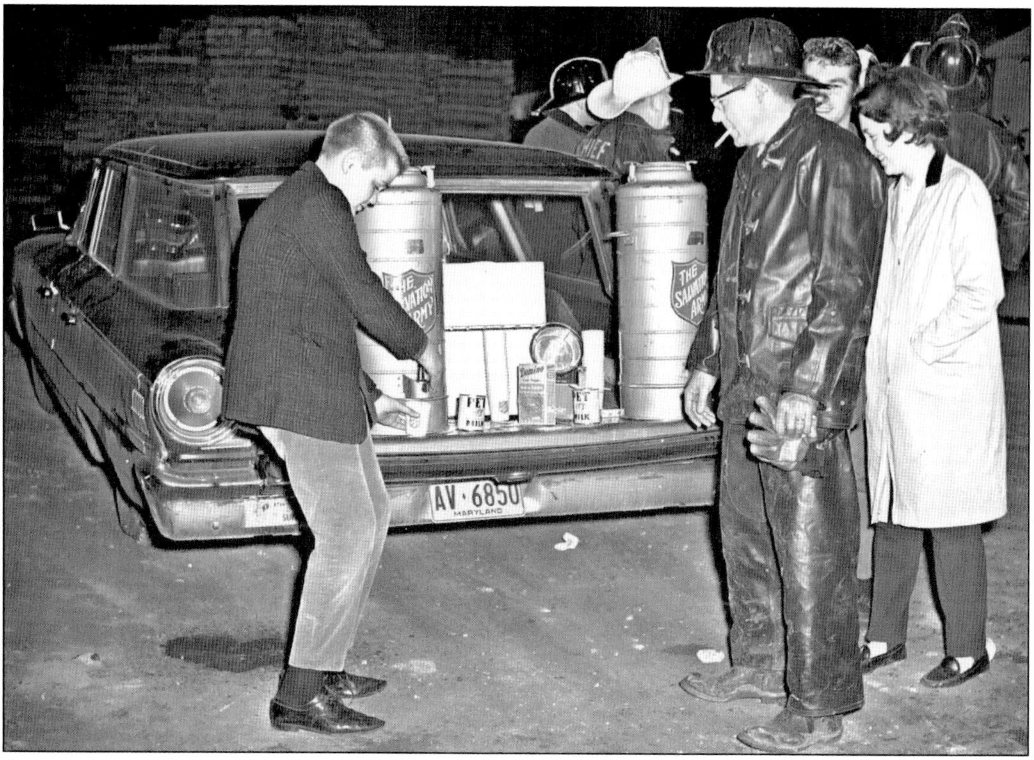

A young man from the Salvation Army pours a cup of coffee for a thirsty firefighter at the Central Chemical Company located on Mitchell Avenue. Fires occurred at Central Chemical quite often. This fire was in the early morning hours of April 10, 1965. (Courtesy of the First Hagerstown Hose Company.)

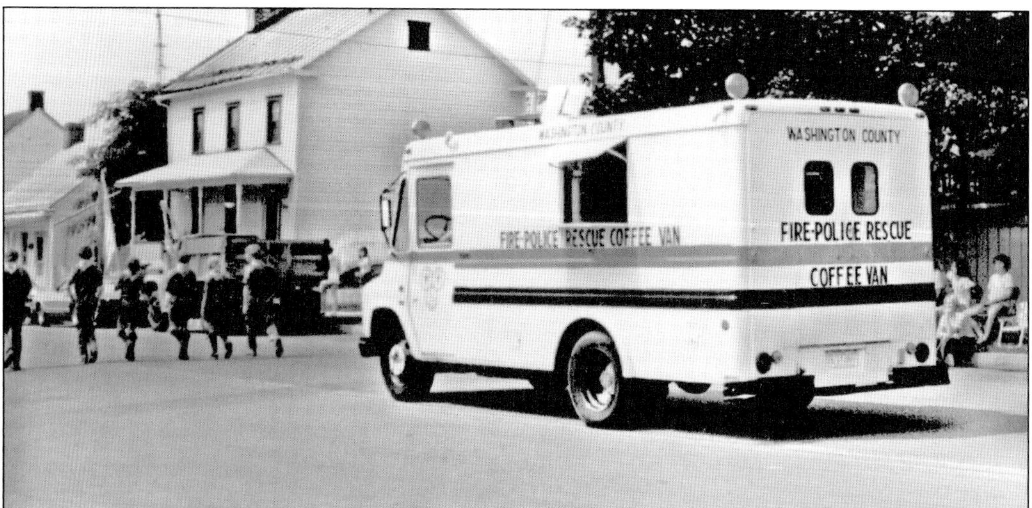

Who takes care of a wet, cold, tired firefighter? The answer is the Washington County Rehab Unit (Unit 255). The Rehab Unit responds to emergency incidents at the request of the officer in charge. Hot and cold food and drinks are provided at no charge to the emergency workers. The photograph was taken at Williamsport, Maryland, during the C&O Canal Days parade in 1977. (Courtesy of the First Hagerstown Hose Company.)

The Washington County Emergency Air Unit (Company 25) was formed in 1966. The truck has an air compressor that is capable of refilling a firefighter's self-contained breathing apparatus cylinder. This roaring fire is in J.J. Newberry's department store located at 75 West Washington Street on March 10, 1974. (Courtesy of the First Hagerstown Hose Company.)

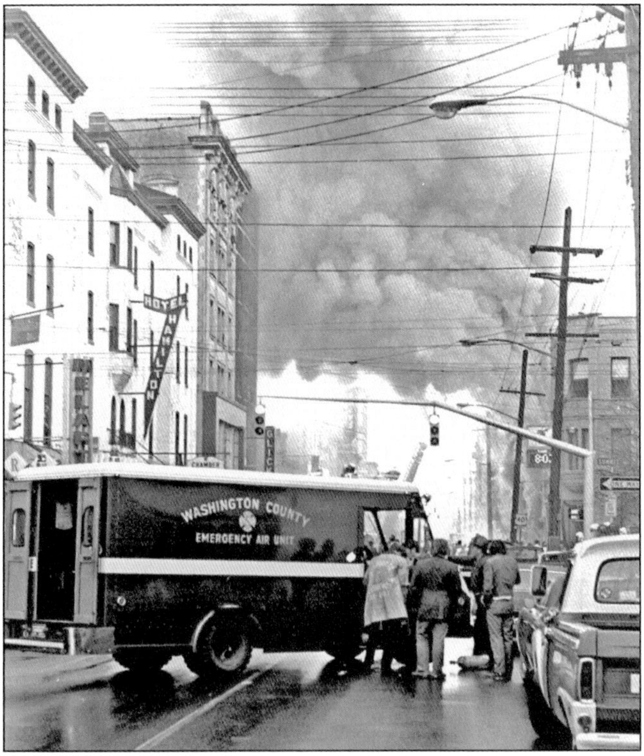

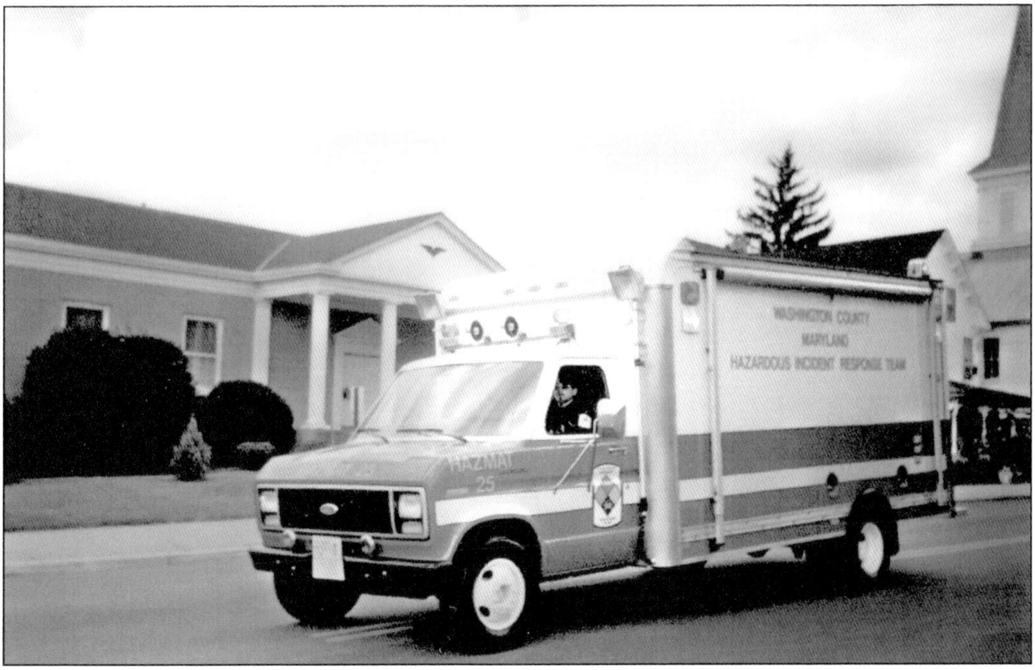

The Washington County Hazardous Materials Response Team (Special Operations 20) was organized in 1985. The Hazardous Materials Response Team provides technical expertise and equipment to handle various emergencies. This is their first response vehicle, a 1986 Ford. (Courtesy of the Warren Jenkins collection.)

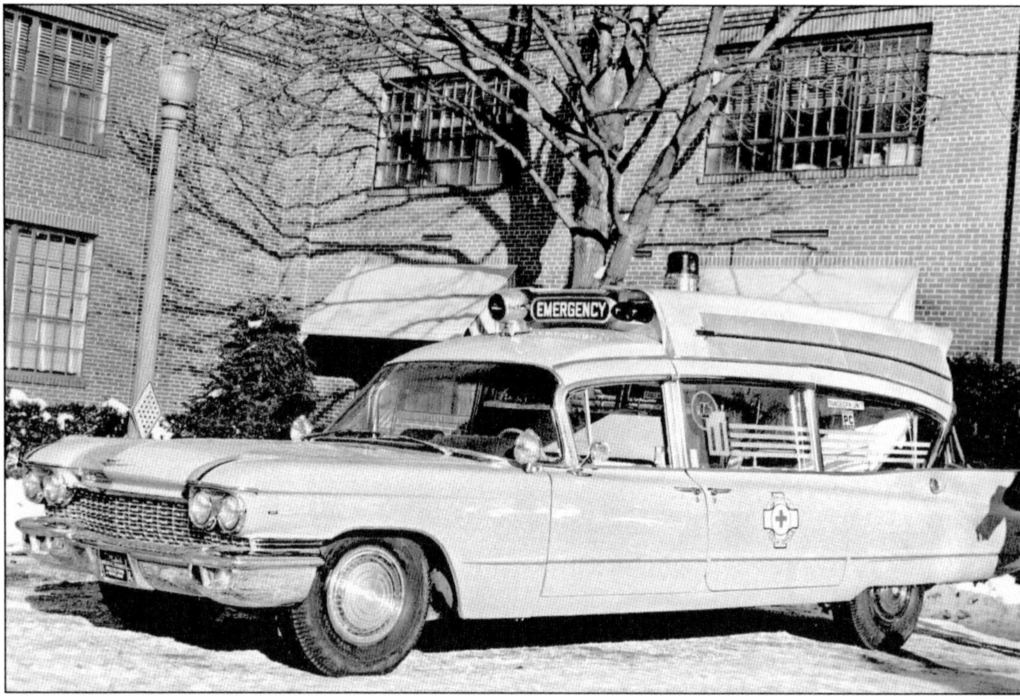

The Community Rescue Service (Company 75) has provided ambulance service to Hagerstown and the surrounding area since September 1955. During a fire in November 1954, a woman jumped from a burning building and injured herself severely. No ambulance was available to handle the emergency, so the Community Rescue Service was born. This ambulance was donated by the Pangborn Corporation. (Courtesy of the Richard Smith collection.)

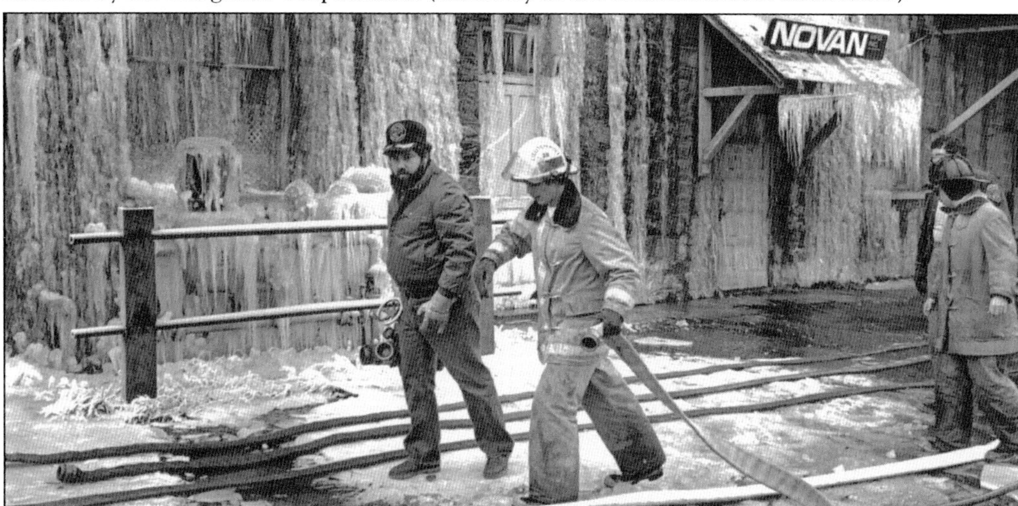

The firefighter on the right is a member of the Washington County Civil Defense (Company 28). This group of dedicated individuals provided support services to the fire department. Personnel and equipment to handle such emergencies as fluid spills, fires, water rescues, and (as in this photograph) flooding conditions were provided. Frozen sprinkler pipes broke at the vacant Novan building located at City Park and created a major flooding condition. (Courtesy of the Todd Gelwicks collection.)

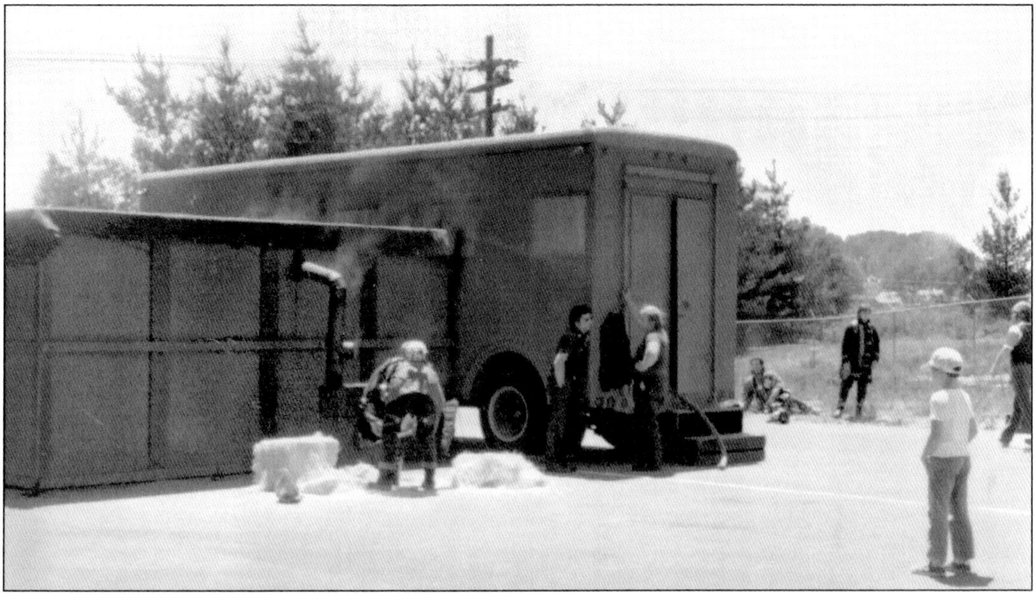

The Maryland Fire and Rescue Institute provides a wide range of training programs to fire/ emergency medical providers across the state. Before a permanent burn building was erected, this mobile teaching lab was used. Bales of excelsior were put into the outside stove to create a near-zero visibility inside. An obstacle course (maze) was also provided. (Courtesy of the Richard Smith collection.)

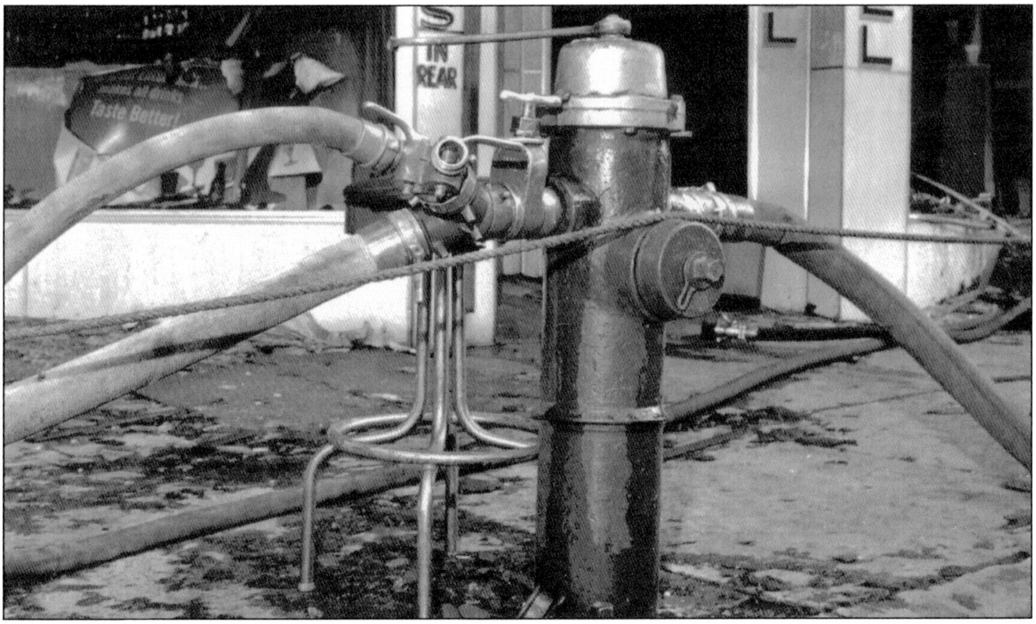

A key component to any fire department is the availability of water. The Hagerstown Water Company, a private venture, began serving Hagerstown in about 1880. The water originated from a reservoir in Smithsburg, Maryland. By 1918, the Hagerstown Board of Water Commissioners managed the system. Two future governors of the state, William Hamilton and Preston Lane, would influence the development of the system. (Courtesy of the First Hagerstown Hose Company.)

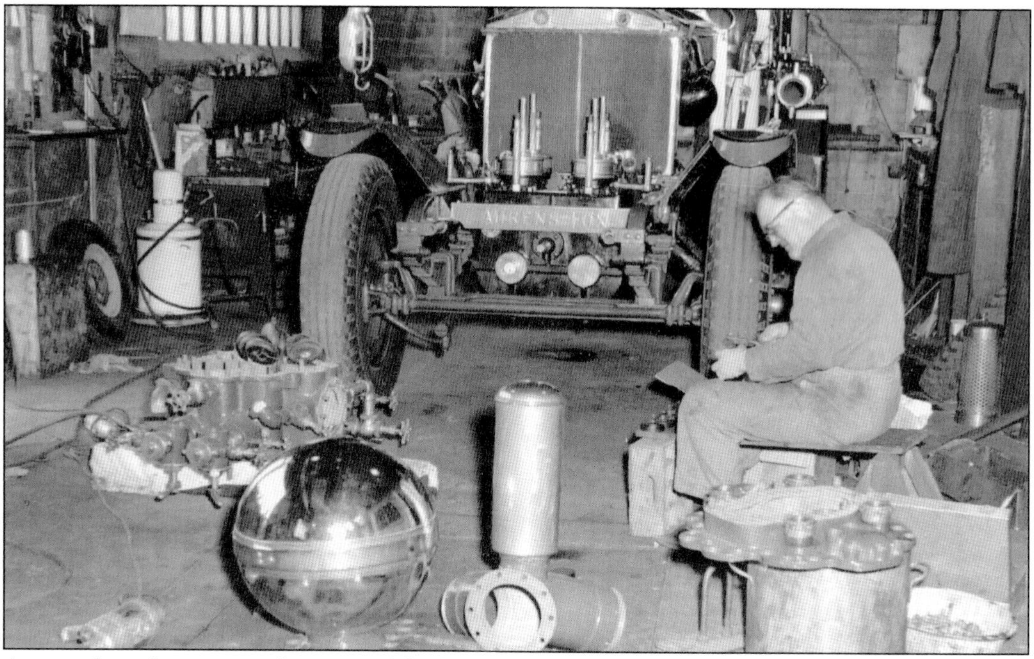

A major but often unseen function of the fire department is the maintenance of the equipment. It is imperative that the pump and mechanical components are ready to handle the next incident. In this photograph, Ahrens-Fox representative Frank Griesser is working on "Dynamite." The work is being performed at the city equipment building located on West Memorial Boulevard. (Courtesy of the C. William Karn collection.)

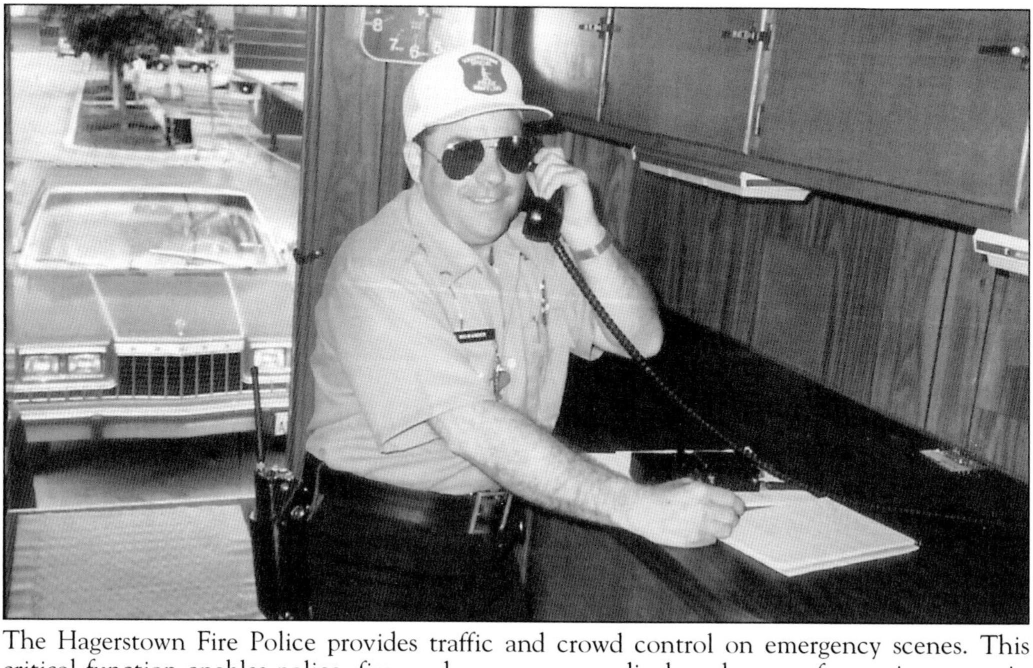

The Hagerstown Fire Police provides traffic and crowd control on emergency scenes. This critical function enables police, fire, and emergency medical workers a safer environment in which to work. Shown in this photograph is a charter member of the organization (which was founded in 1983), Marvin Niswander. (Courtesy of the Marvin Niswander collection.)

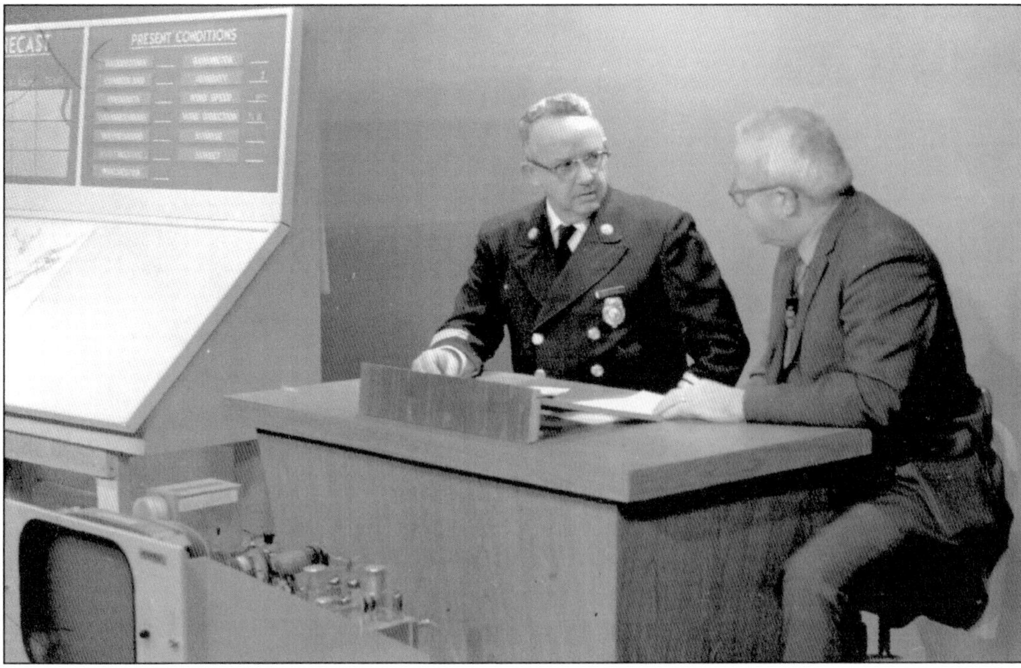

The media provides an important role to the fire department. Stories of fires and other emergencies provide the public with critical information relative to their safety. Here we see Chief Hall (left) being interviewed by longtime WJEJ radio personality Ed Klitch. (Courtesy of the Richard Smith collection.)

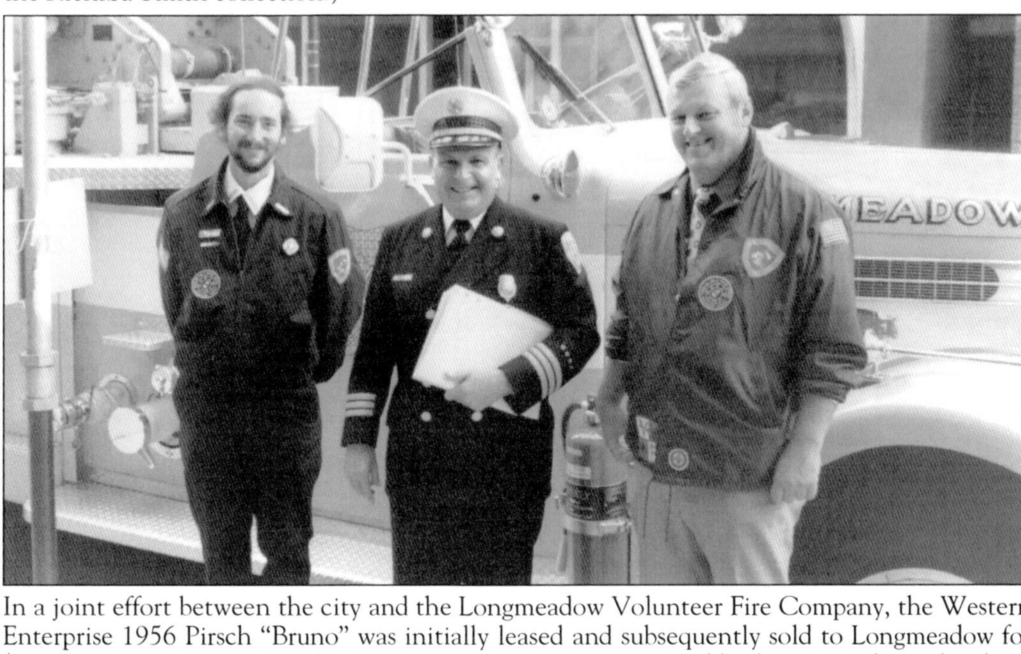

In a joint effort between the city and the Longmeadow Volunteer Fire Company, the Western Enterprise 1956 Pirsch "Bruno" was initially leased and subsequently sold to Longmeadow for $1. Keeping Bruno close to the city ensured its ability to respond back in town for multi-alarm fires. Longmeadow Fire Chief Terry Burkholder (left) and Dep. Chief C. Michael Harshman (right) are shown here accepting the truck from Chief Richard Smith (center). (Courtesy of the Richard Smith collection.)

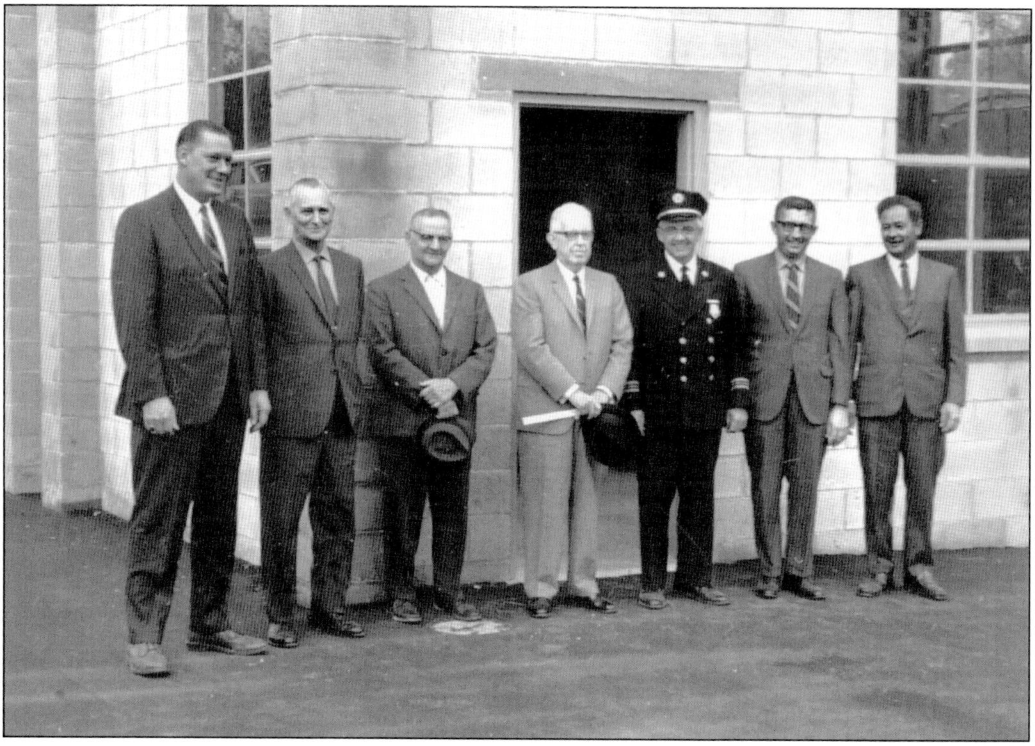

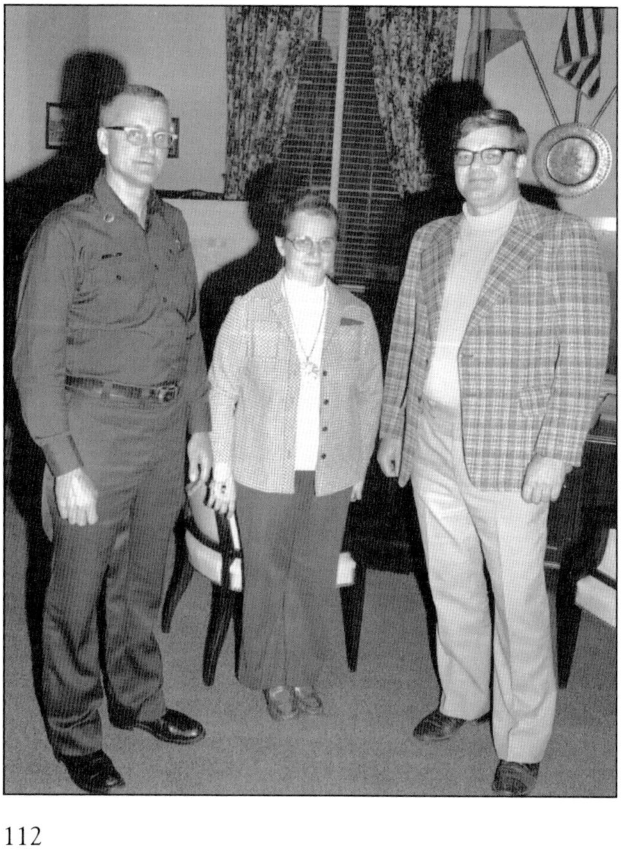

It is vital that the mayor, council, and the fire department work closely together to provide the best possible fire protection to the citizens. Here, Mayor Mills (center), Chief Hall, and other city officials are dedicating the Hagerstown Training Center on October 5, 1969. (Courtesy of the Antietam Fire Company.)

Being a good firefighter requires dedication and sacrifice. A firefighter may be called to a fire at any moment day or night, including holidays. A firefighter's family endures these hardships also. It is often a firefighter's wife or husband who maintains some kind of family normalcy. A firefighter's hero is his or her family. In this photograph, FAO Charles Baker, with his wife Janice, is being promoted to the rank of lieutenant by Mayor Varner Paddack. (Courtesy of the Richard Smith collection.)

Ten

ON THE JOB

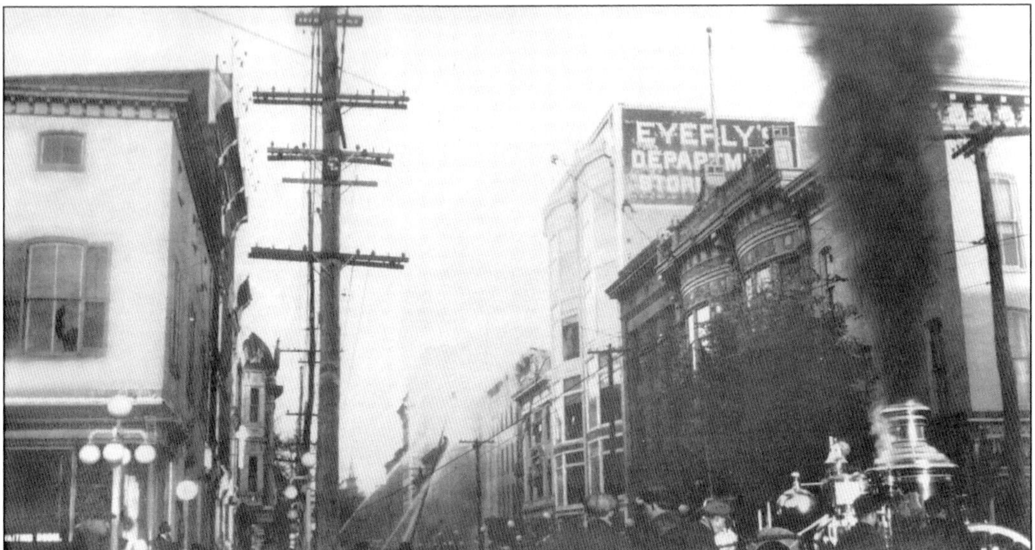

It is believed that embers from the Antietam Garage and Antietam Fire Company fire on Summit Avenue set fire to the Baldwin Hotel located in the unit block of West Washington Street on October 28, 1914. Note the steamer pumping in the public square and the multiple fire streams being directed towards the fire. (Courtesy of the First Hagerstown Hose Company.)

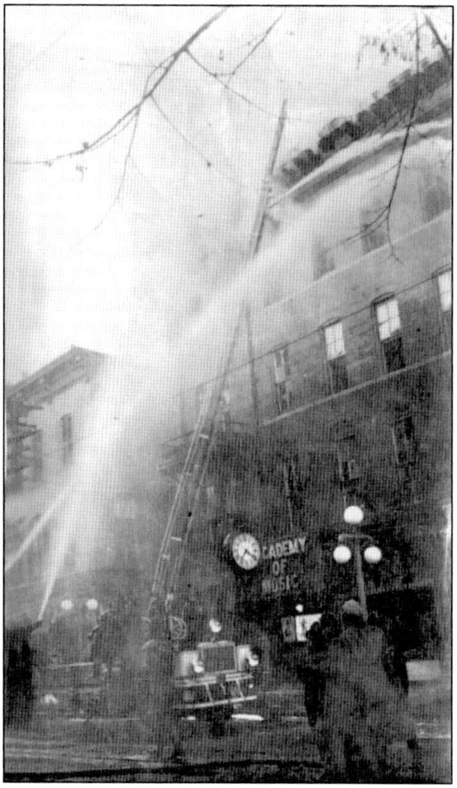

This fire at the Baldwin Hotel is the site of the Washington House (hotel) fire in 1879. Note the Pioneer Hook & Ladder's 1914 American LaFrance directing a fire stream (ladder pipe operation) onto the burning building. (Courtesy of the First Hagerstown Hose Company.)

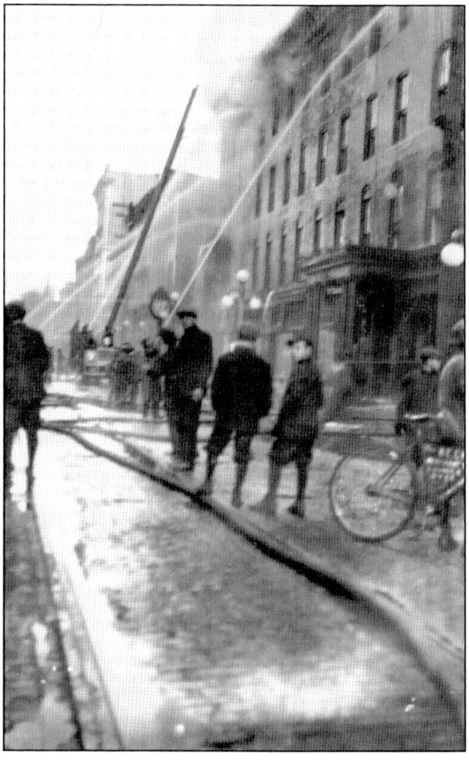

This is another view of the Baldwin Hotel fire. These are some of the earliest known photographs of the Hagerstown Fire Department in action. Note the large crowd of all ages who have gathered to watch the spectacle. (Courtesy of the First Hagerstown Hose Company.)

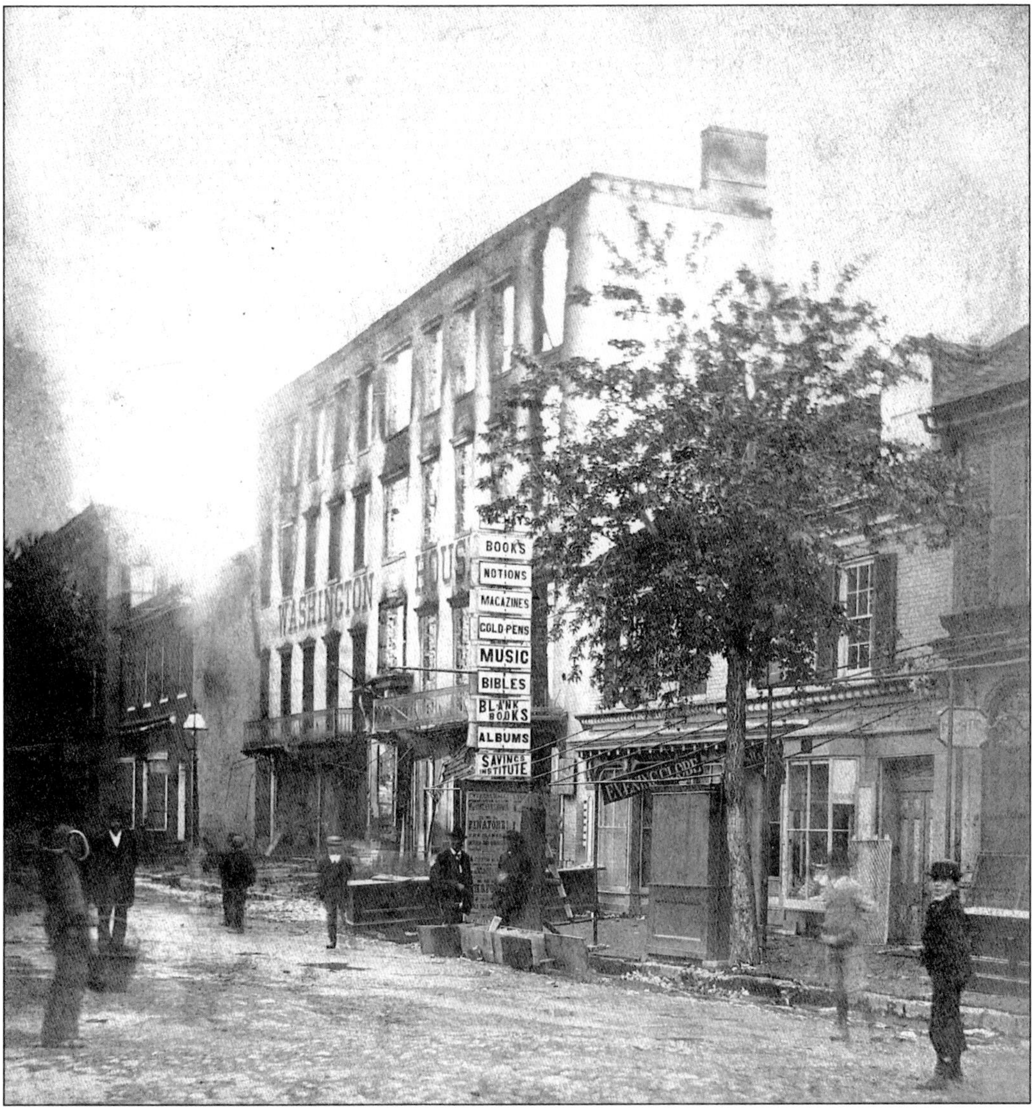

On a beautiful, bright calm night in June 1879, a fire was discovered under the first-floor steps of the Washington House (hotel) located in the unit block of West Washington Street. Of the 65 guests, one perished and many were injured when the fire cut off their escape route. Firefighters rescued several citizens via ladders. Pioneer Hook & Ladder firefighter Samuel Dorsey rescued two women and a child from the third floor. Upon hearing their cry to be saved, Dorsey entered the building and began a search. When he found the women and child he wrapped them with bed clothing and even wrapped his own coat around the child. At first he was driven back by the smoke and heat but he finally reached the street. His hands and face were seriously burned and his hair singed, but the women and child were safe. (Courtesy of the Washington County Historical Society.)

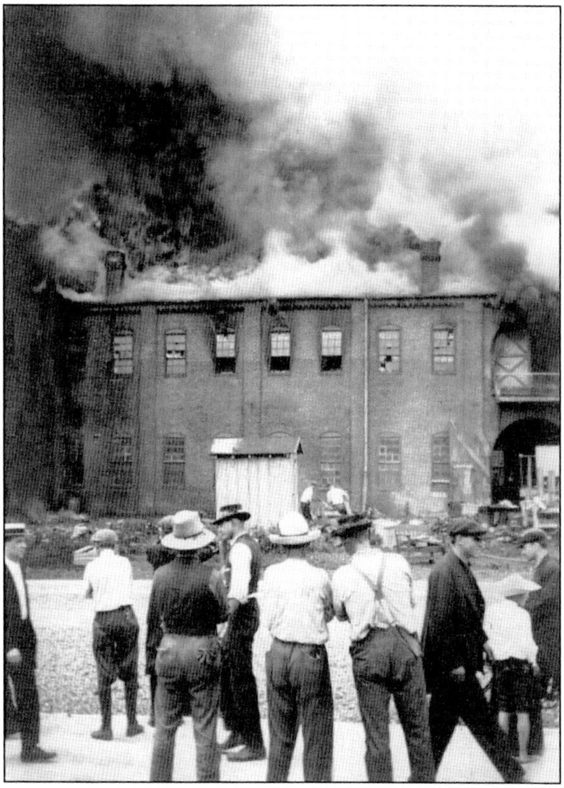

It was not uncommon for major fires to occur in manufacturing buildings in late 1800s and early 1900s. Very few safety standards existed. This major fire occurred at the Steam Engine Works at noontime on Wednesday, June 16, 1915. (Courtesy of the First Hagerstown Hose Company.)

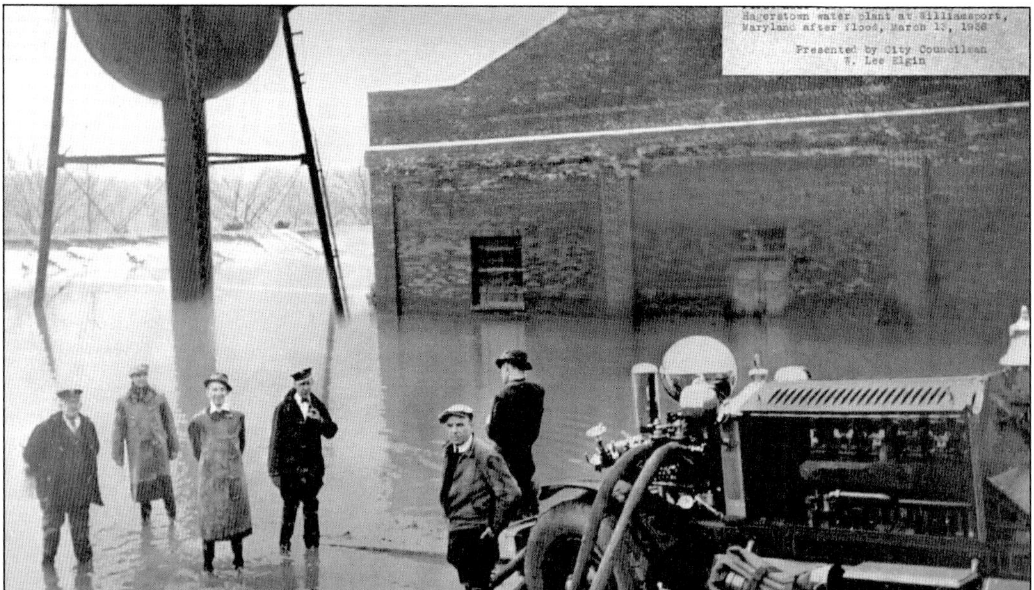

One of the largest floods to ever strike this region occurred in 1936. In this photograph, the hose company's 1923 Ahrens-Fox "Dynamite" is pumping water from the Hagerstown Water Plant located in Williamsport, Maryland. The Potomac River is flowing through the plant. (Courtesy of the First Hagerstown Hose Company.)

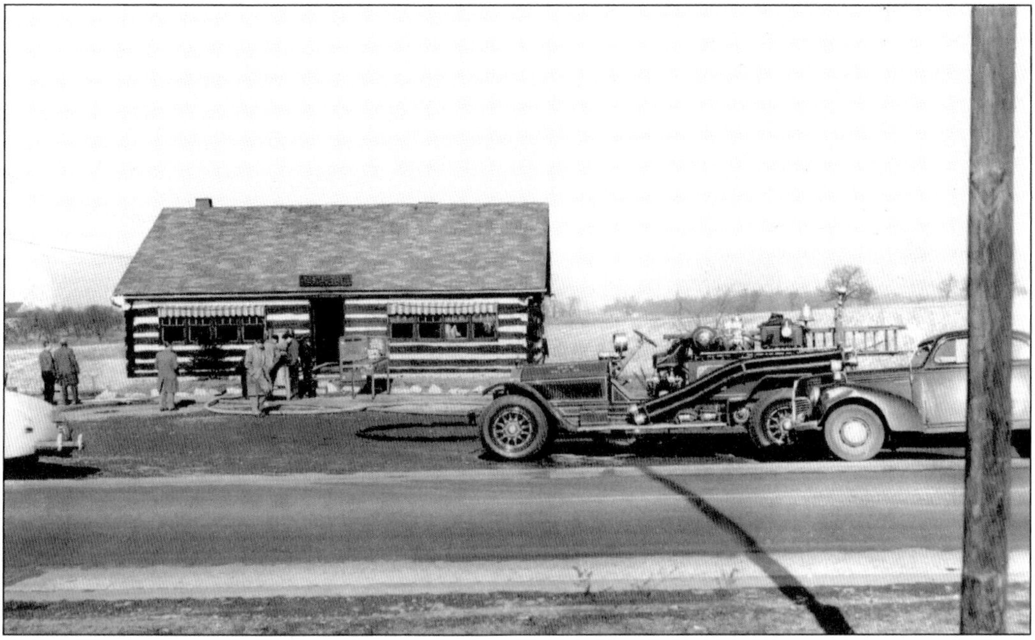

The Western Enterprise Fire Company's 1925 American LaFrance "Maud" is shown on the scene of the Knotty Pines Restaurant fire, located on West Washington Street Extended (Route 144). The picture was taken c. 1947. A more serious fire would destroy the building in the future. This is a rare photograph of Maud in action. (Courtesy of the William Dieterich collection.)

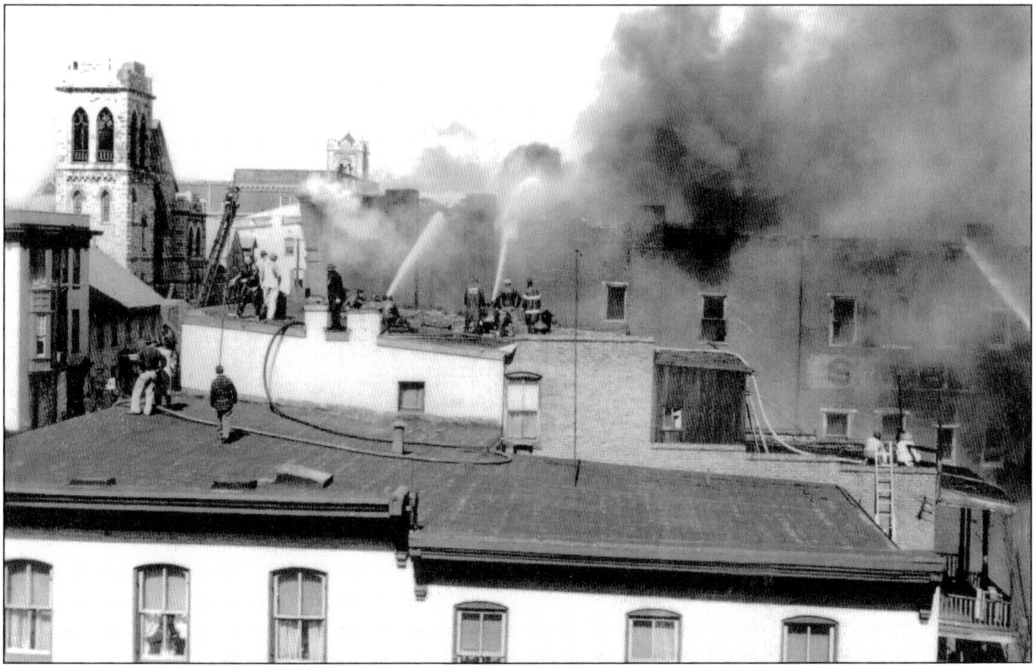

Firefighters are dousing the flames from the rooftops of nearby buildings to contain this multi-alarm fire. The fire is burning in the S.M. Bloom building located in the 100-block of North Potomac Street on March 24, 1948. (Courtesy of the William Dieterich collection.)

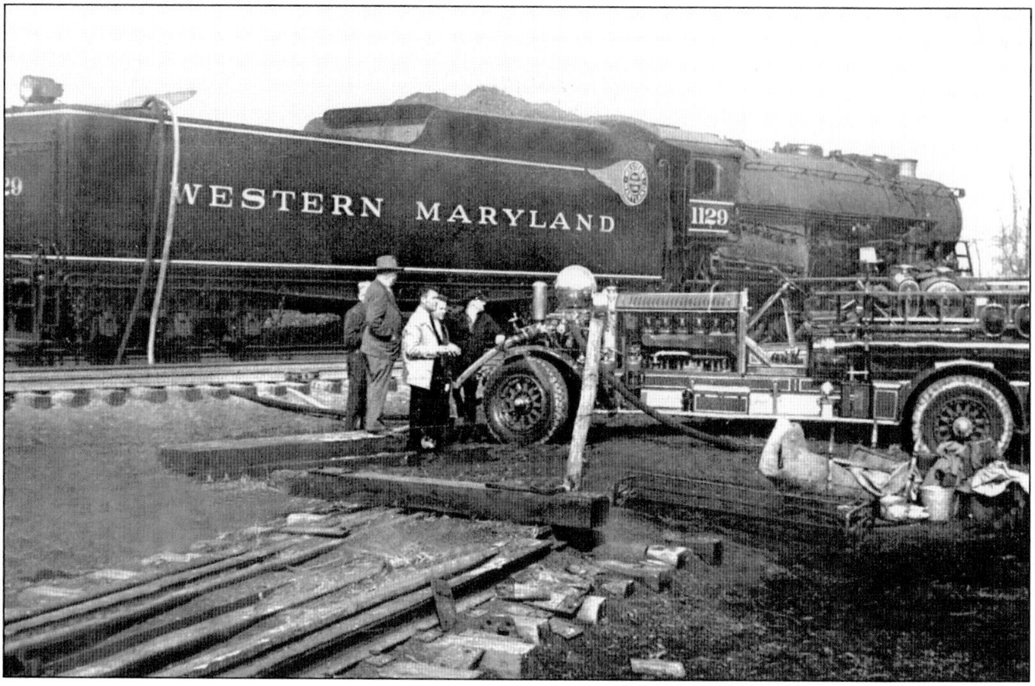

There is more than one way to get water. Here, water for the old Marl Plant fire is obtained from this railroad water tender. The Marl Plant was located off Hopewell Road; marl is a form of lime used for fertilizer. The Antietam's "Queen Mary" is pumping water to the fire. From left to right are unidentified, George Harper, Driver Kenny Myers, John Shifler, and Driver Ralph Foltz. (Courtesy of the William Dieterich collection.)

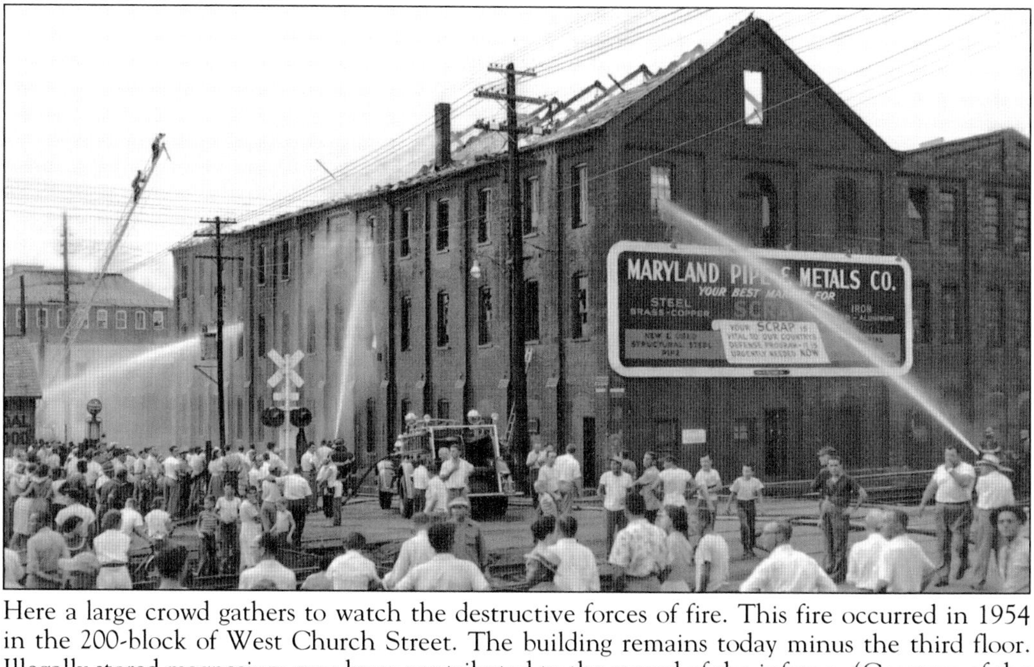

Here a large crowd gathers to watch the destructive forces of fire. This fire occurred in 1954 in the 200-block of West Church Street. The building remains today minus the third floor. Illegally stored magnesium may have contributed to the spread of the inferno. (Courtesy of the John Hall collection.)

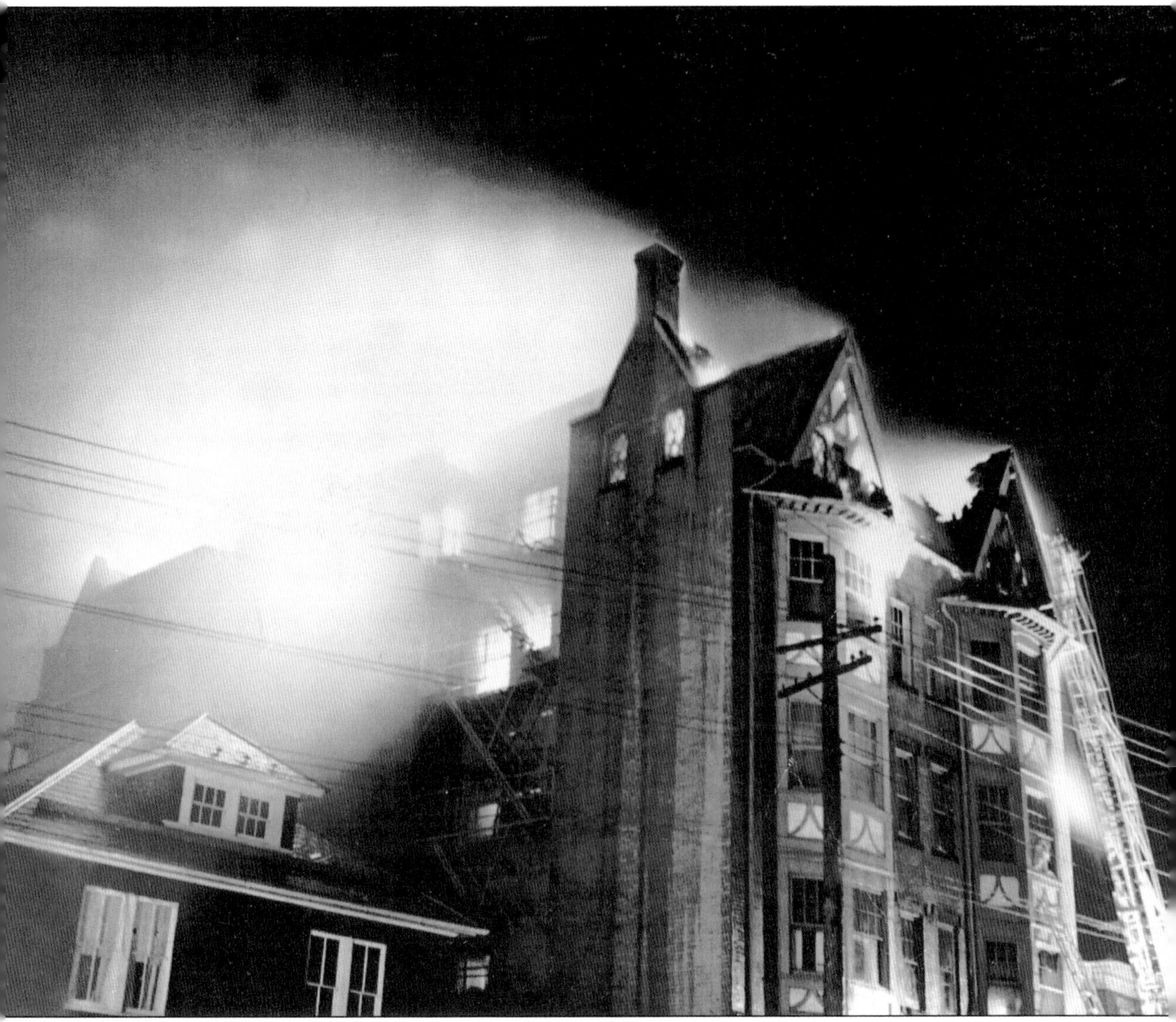

Those who were there will never forget this fire. On the very cold winter night of December 30, 1955, fire broke out in the Grey Gables Apartment building located at 600 Potomac Avenue. An elderly female was rescued from a front window by firefighters. The woman found the window by following the grooves in the floorboard. The building is still an apartment building minus the top two floors. (Courtesy of the William Dieterich collection.)

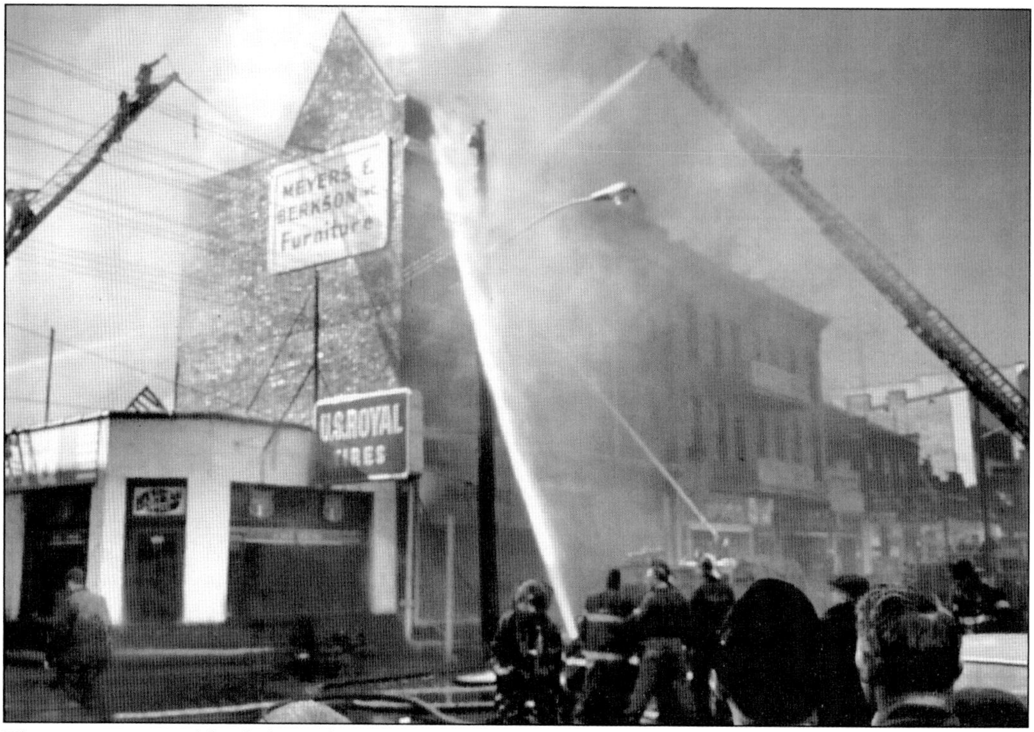

This is a picture of firefighters battling flames at the Myers/Berkson Furniture Store fire, located in the unit block of West Franklin Street, on January 9, 1963. A furniture store fire is one of the most dangerous to fight. The showroom is usually a maze of rooms, an easy place in which a firefighter could get lost. The furniture itself presents a high fuel-load hazard. (Courtesy of the First Hagerstown Hose Company.)

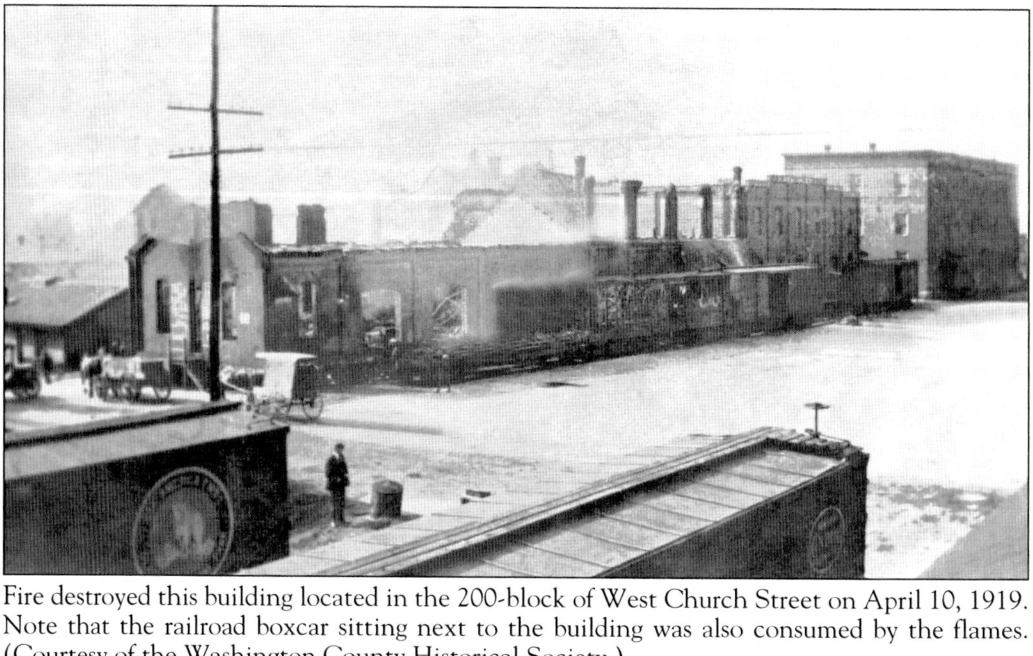

Fire destroyed this building located in the 200-block of West Church Street on April 10, 1919. Note that the railroad boxcar sitting next to the building was also consumed by the flames. (Courtesy of the Washington County Historical Society.)

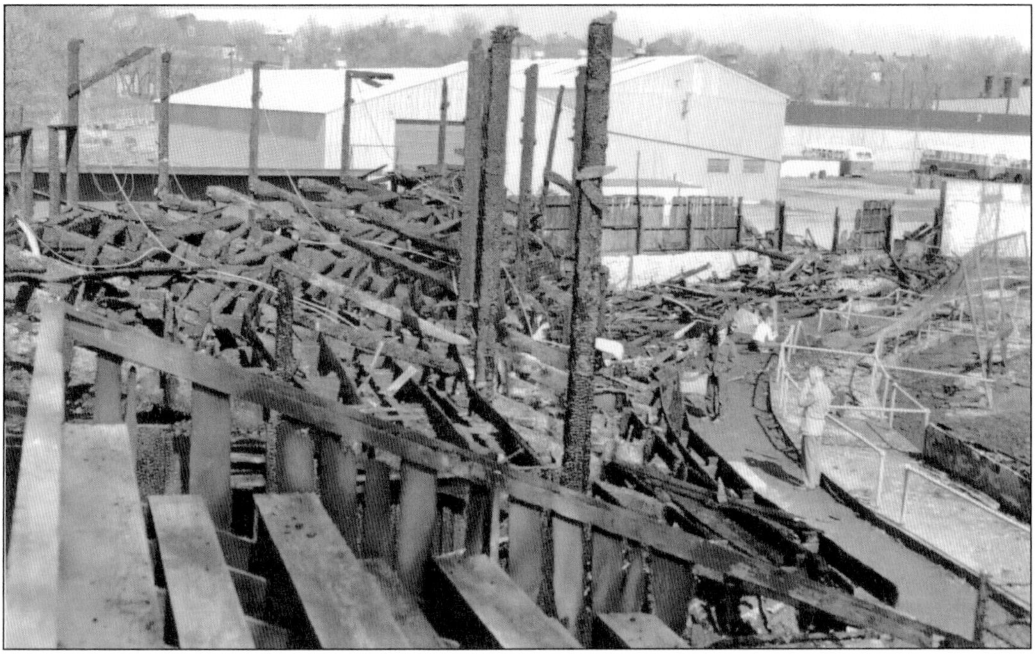

This is all that remains of the wooden seats at Municipal Stadium on April 29, 1961. The stadium has been called home by numerous professional and semi-professional baseball teams, including the Hagerstown Owls, the Hagerstown Braves, and in the past 20 years the Hagerstown Suns. (Courtesy of the First Hagerstown Hose Company.)

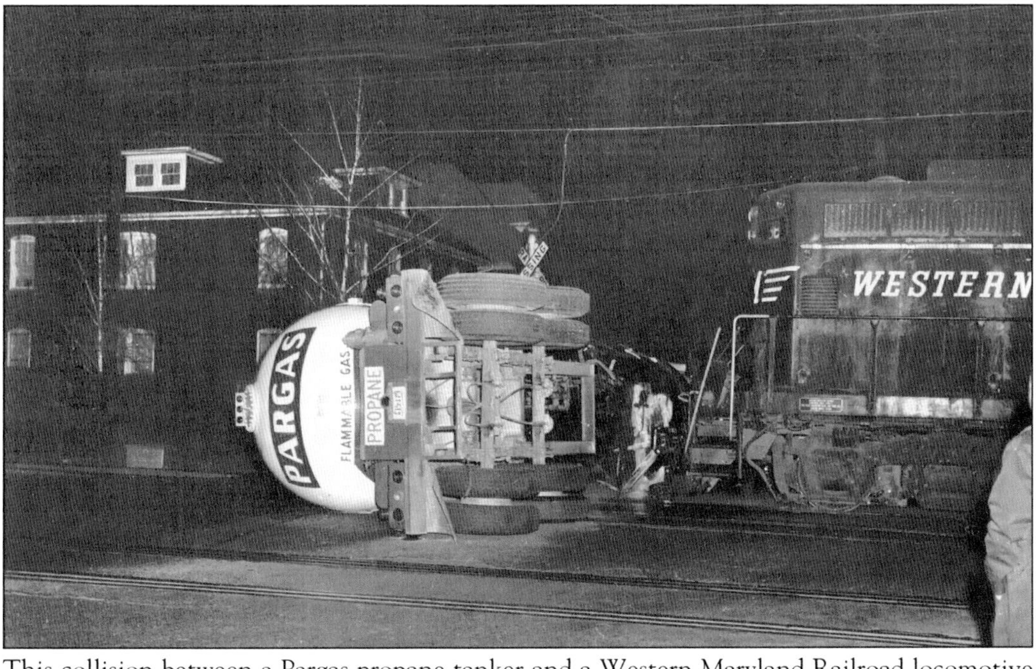

This collision between a Pargas propane tanker and a Western Maryland Railroad locomotive occurred on February 27, 1968. As a result of this very volatile situation, several blocks in all directions were evacuated. Fortunately, the incident occurred without an explosion. (Courtesy of the First Hagerstown Hose Company.)

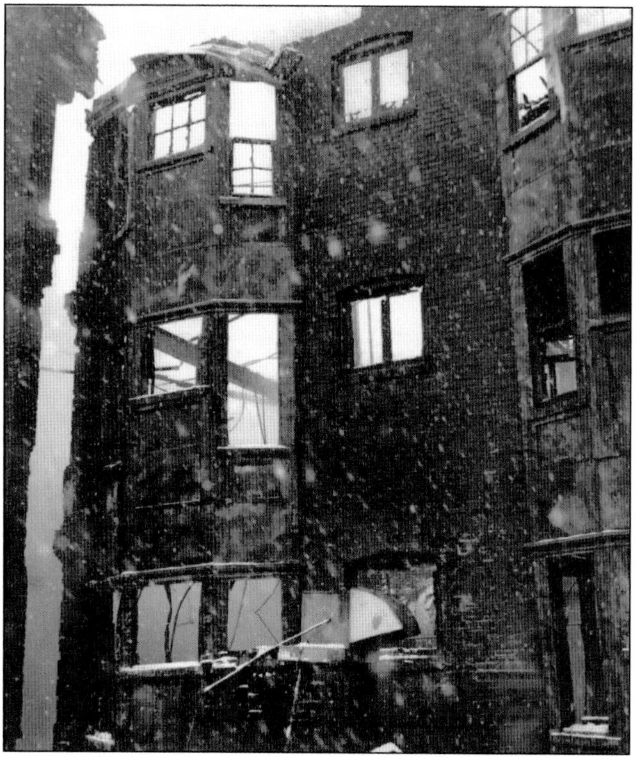

This is the aftermath of the Maryland Apartment fire located at 21 South Potomac Street. At about 1:30 a.m. on the cold winter morning of February 8, 1974, box alarm 1–1–1 (at the public square) was sounded. As the firefighters were dousing the flames, the marquee collapsed, pinning several firefighters for an hour. The firefighters were treated at the hospital for non–life-threatening injuries. (Courtesy of the John Hersh collection.)

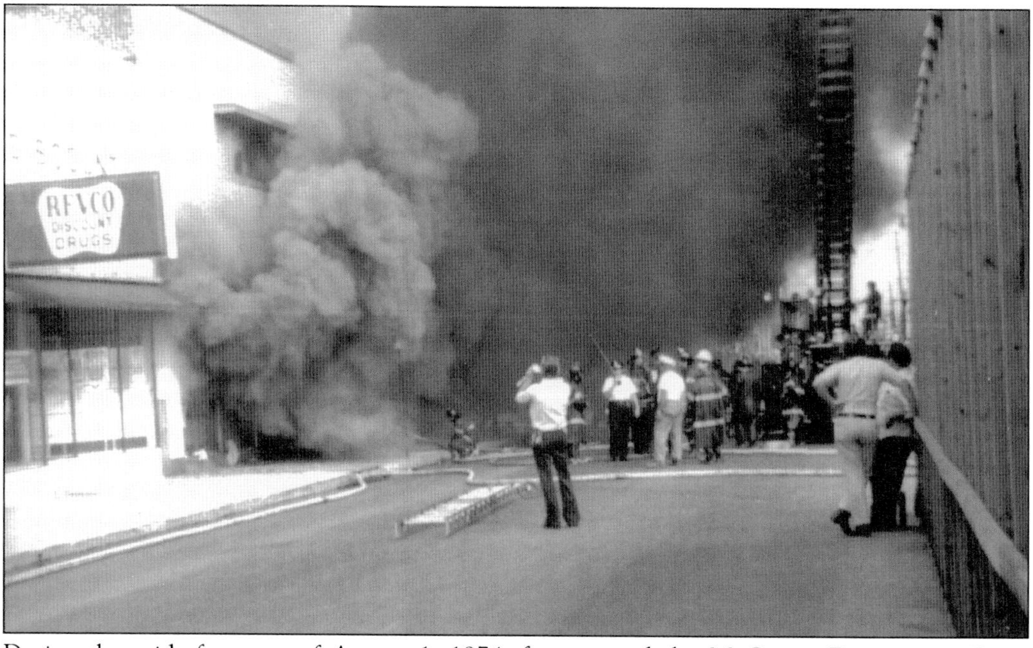

During the mid-afternoon of August 1, 1974, fire ravaged the McCrorys Department Store located at 52 West Washington Street. The estimated loss was $645,000. It's notable that this was the fire department's first shift after switching from 24-hour shifts to 10- to 14-hour shifts. From this day forward there would be three shifts, each commanded by a battalion chief. (Copyright *The Herald-Mail* Company, Hagerstown, Maryland.)

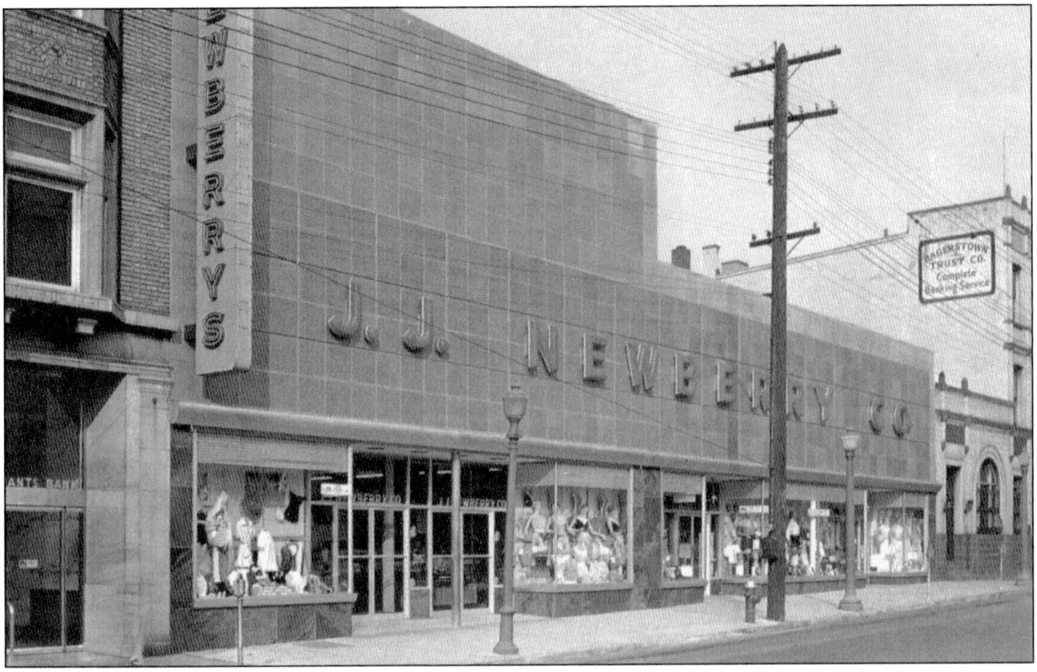

The J.J. Newberry store contained over 30,000 items on three large selling floors and more than a mile of selling space. It also contained a lunch counter and snack bar. (Courtesy of the Richard Smith collection.)

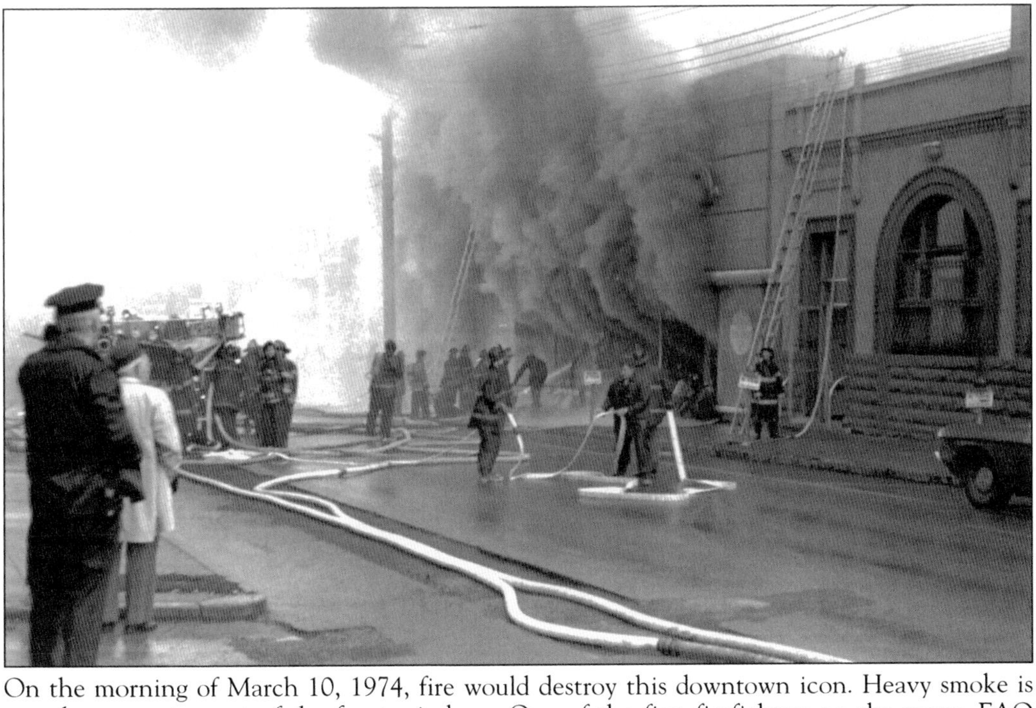

On the morning of March 10, 1974, fire would destroy this downtown icon. Heavy smoke is seen here pouring out of the front windows. One of the first firefighters on the scene, FAO Ben Conrad, thought the manikins lying in the street were real humans. (Courtesy of the First Hagerstown Hose Company.)

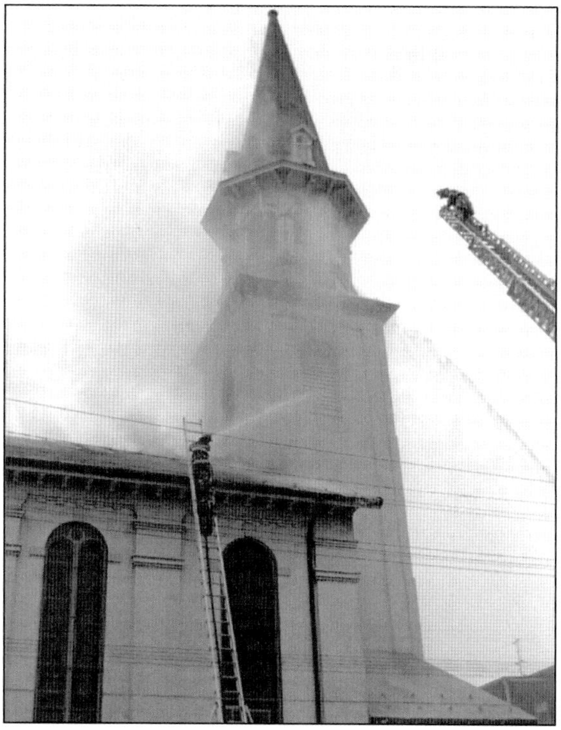

A lightning strike during a thunderstorm was the cause of this fire at Saint John's Lutheran Church, located at 141 South Potomac Street, on July 20, 1969. Firefighters performed exceedingly well at containing the fire to the church tower. (Courtesy of the First Hagerstown Hose Company.)

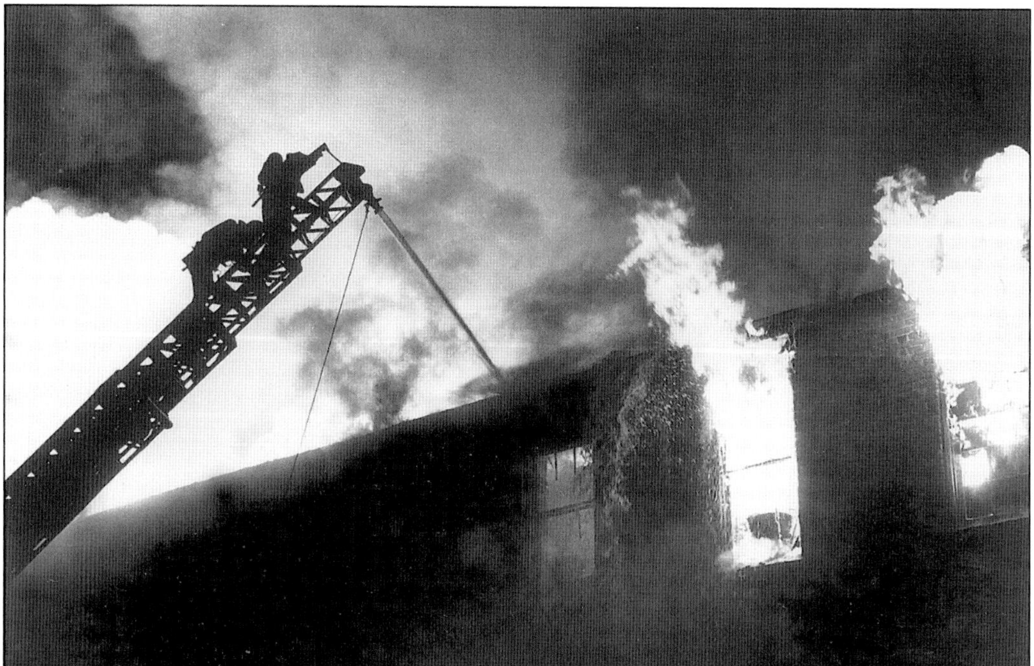

Pioneer Hook & Ladder firefighters have the ladder pipe in service at the Style-Rite Furniture store fire, located at 225 South Mulberry Street, on January 10, 1975. Approximately 125 firefighters fought this five-alarm fire. (Copyright *The Herald-Mail* Company, Hagerstown, Maryland.)

Part of a Hagerstown landmark was destroyed on November 24, 1987, when the dome and top floor of the Kohler building was destroyed by fire. The Kohler building is located at 27 West Washington Street and was constructed in 1880. This aerial view shows the potential this fire had of spreading to nearby buildings. (Copyright *The Herald-Mail* Company, Hagerstown, Maryland.)

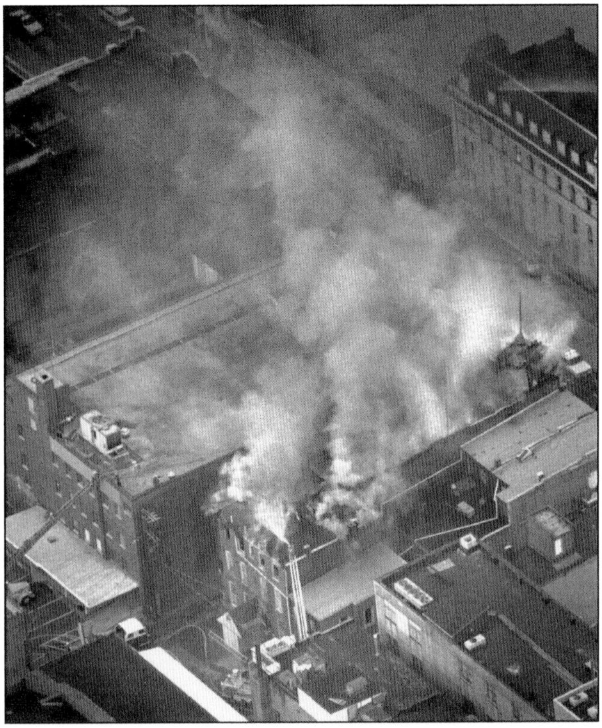

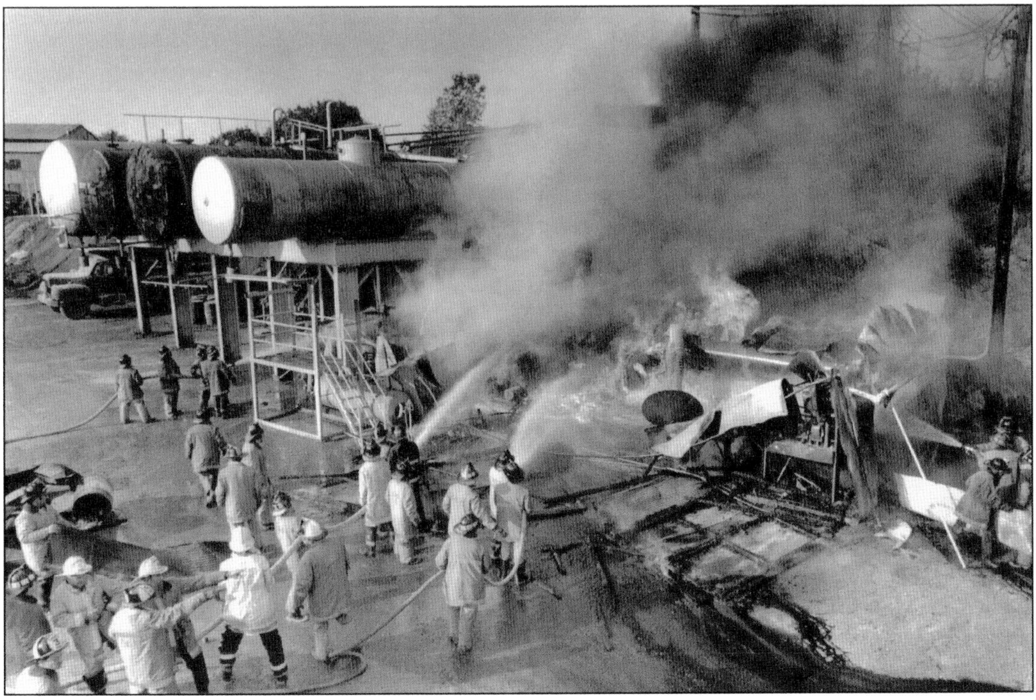

A fire at the Kline Asphalt plant on June 23, 1981, caused déjà vu for Batt. Chief C. Kingsley Poole. Five years earlier almost to the day, a similar fire destroyed the plant. The tanks contained naphtha, oil, and kerosene. High-tension and natural-gas lines were also a major concern. (Copyright *The Herald-Mail* Company, Hagerstown, Maryland.)

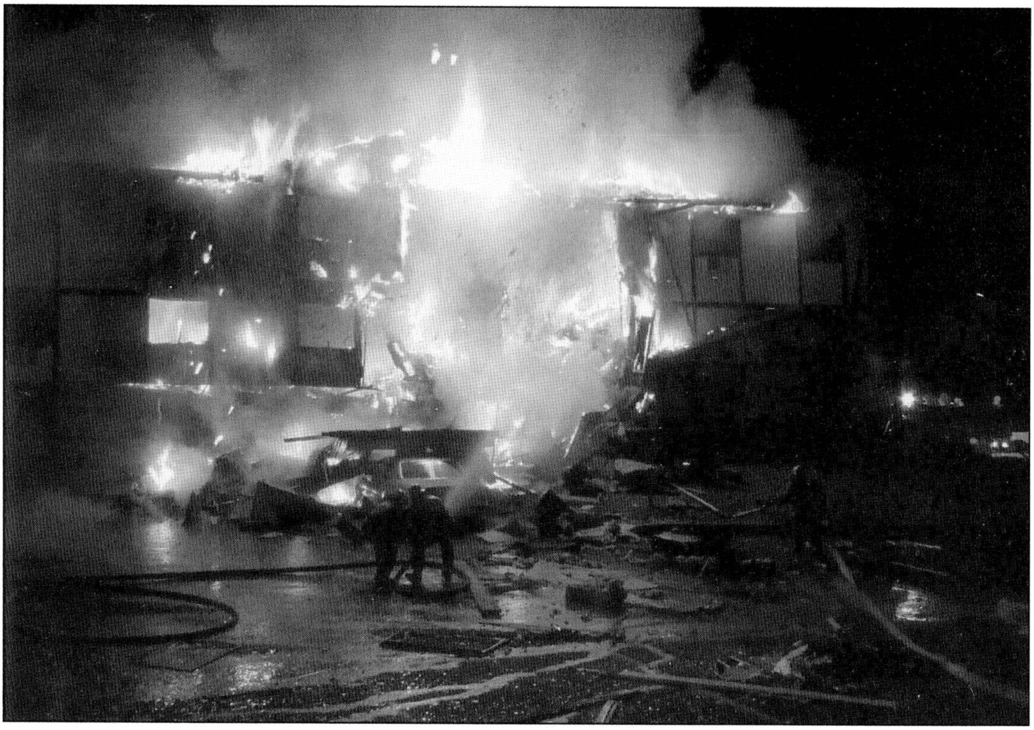

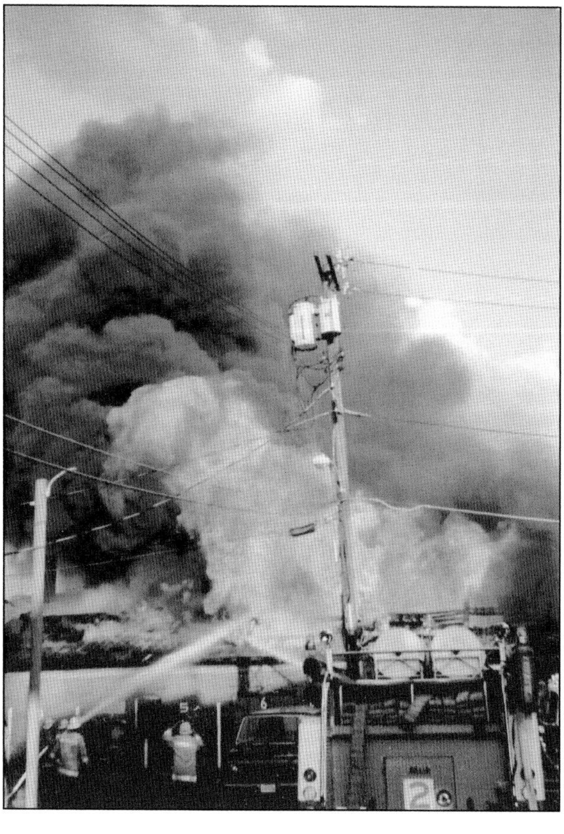

Four people perished and ten were injured on February 18, 1990, at the Super 8 Motel explosion/fire located at 1220 Dual Highway. The cause of the explosion was attributed to a natural-gas leak. Numerous people were rescued by firefighters via ladders because the explosion had jammed their doors. Occupants related the explosion to an earthquake. (Courtesy of the Tom Brown collection.)

The Leeds Men's Wear Store was a family business for 47 years. On the evening of April 10, 1991, fire destroyed the building. With a fierce wind hampering firefighting efforts, the fire spread to an adjoining building, the Southwest Trading Company restaurant. Smoke could be seen over most of the east end of town throughout the evening. (Courtesy of the Tom Brown collection.)

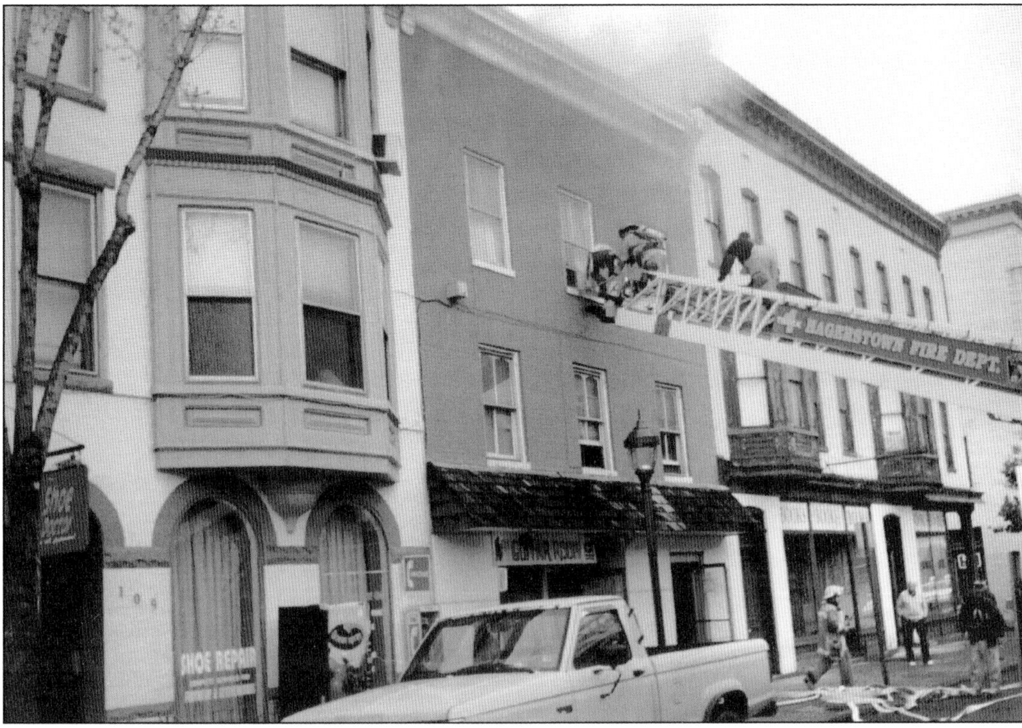

On April 7, 2001, fire ripped through this apartment building located at 105 South Potomac Street. In this rare glimpse of an actual rescue are FAO Andrew Hartman and Firefighter Tony Lida helping a man down his only escape route, the aerial ladder. (Courtesy of the author.)

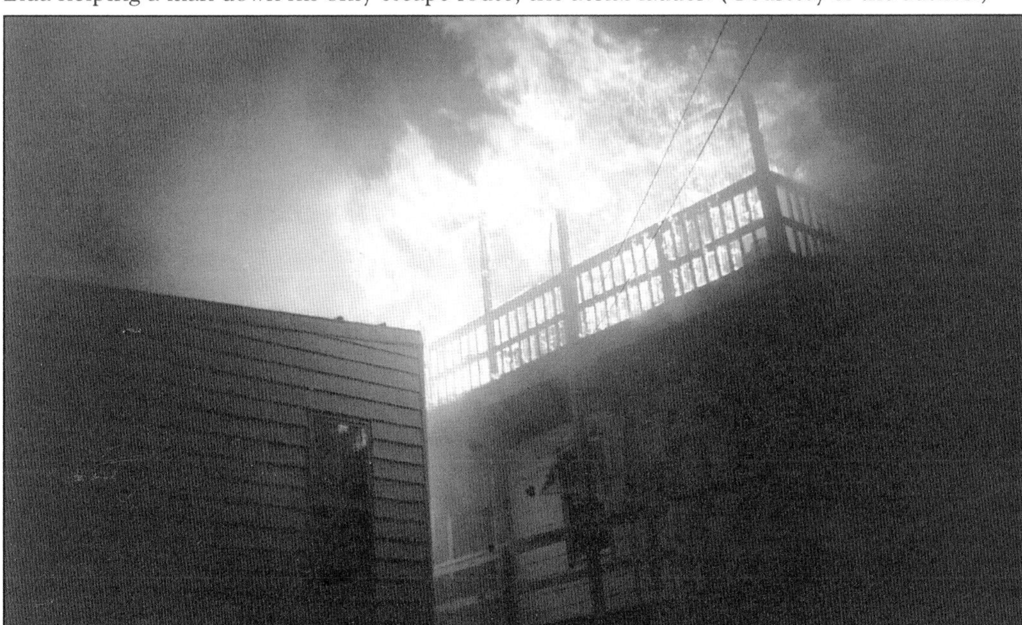

Downtown fires that are in close proximity to other buildings can create a non-containment situation quickly. This apartment fire at 38–40 Randolph Avenue on the night of May 19, 2002, had such a potential. (Courtesy of the author.)

ACKNOWLEDGMENTS

What started out as a vision by one person quickly became a community-wide project. Upon a plea for photographs and historical documents, many people came forward to share their collections, which the author gratefully acknowledges.

It seems that as a young boy William Dieterich had the forethought of a grown man. Bill took full advantage of the opportunity to speak to the drivers of the horse-drawn days. Bill remembered those stories and would later become a legendary firefighter in his own right.

This project would not be possible without the support of the management of the Hagerstown Fire Department. Chief Gary Hawbaker and Acting Chief (now retired) Richard Kipe offered full cooperation throughout the process.

The following individuals and organizations offered photographs or historical information: First Hagerstown Hose Company, Antietam Fire Company, Independent Junior Fire Company, Western Enterprise Fire Company, South Hagerstown Fire Company, Pioneer Hook & Ladder Company, Mindy Marsden and the Washington County Historical Society, John Frye and the Western Maryland Room of the Washington County Free Library, Washington County Fire and Rescue Communications Center, Hagerstown Engineering Department, Hagerstown Mayor William Briechner and First Lady Gann Briechner, retired Fire Chief John Hall, retired Fire Chief Richard Smith, Batt. Chief Kyd Dieterich, Blaine Snyder, Frank Ridenour, Terry Snider, Richard Stallings, James Sprecher Jr., William Dunham, Jeff Snyder, Ryan Evans, Richard Rohrer, James Schaffer, Terry Mose, Jerry Cunningham, Rodney McCoy, James Plummer, Richard Conrad, Fire Marshal Tom Brown, Fire Marshal John Hersh (retired), Michael Churchy, Robert S. Kefauver, Lucy Karn, Mable Basore, JoAnn Logan, David White, Warren Jenkins, Brenda Hawbaker, and Charles Shindle.

Joe Goodrich provided technical assistance. Thanks to the Mayhue family, wife Diana, daughter Lisa, and son Philip, for your patience and support.

The photographer is unknown for numerous photographs but your contribution is appreciated.

Reference books included the following: *Chemical Fire Engines* by W. Fred Conway (1987); *The Encyclopedia of American LaFrance* by Robert B. Marvin, D.P.A. (1995); *The Fireman* by the American Association of Firemen (1928); *History of Western Maryland* by J. Thomas Scharf, A.M. (1968); *A History of Washington County, Maryland* by Thomas J.C. Williams (1906); "Maryland State Firemen's Convention" booklet (1936); *Ahrens-Fox: The Rolls-Royce of Fire Engines* by Ed Hass (1982); "25th Anniversary of IAFF Local 1605" booklet (1991).

Thanks to the following local media for sharing this project with the community: WHAG, TV-25 news director Mark Kraham, *The Herald Mail* newspaper publisher John League, veteran reporter Marlo Barnhart, and photography director Kevin Gilbert.